T0092827

Problem-Solving

Problem-Solving
Leaning on New Thinking Skills

Howard Eisner

CRC Press
Taylor & Francis Group
Boca Raton London New York

CRC Press is an imprint of the
Taylor & Francis Group, an **informa** business

First edition published 2021
by CRC Press
6000 Broken Sound Parkway NW, Suite 300, Boca Raton, FL 33487-2742

and by CRC Press
2 Park Square, Milton Park, Abingdon, Oxon, OX14 4RN

Library of Congress Cataloging-in-Publication Data
Names: Eisner, Howard, 1935- author.
Title: Problem-solving : leaning on new thinking skills / Howard Eisner.
Description: First edition. | Boca Raton, FL : CRC Press, 2021. | Includes bibliographical references and index. | Summary: "This book presents various methods of problem-solving that can be adapted to any field. It focuses on a set of a dozen new approaches with an ending result to finding better solutions to problems that you may have previously found difficult. The book discusses problem-solving based upon new thinking skills and presents the relationship between problem-solving and creativity. A connection between problem-solving and re-engineering is presented as the book explores the ability to tackle new and difficult problems in all aspects of life. Target audience is general engineers, systems engineers, scientists, technologists, mathematicians, and lawyers"— Provided by publisher.
Identifiers: LCCN 2021004488 | ISBN 9780367749811 (hbk) | ISBN 9780367749828 (pbk) | ISBN 9781003160618 (ebk)
Subjects: LCSH: Problem solving. | Creative thinking.
Classification: LCC T57.95 .E43 2021 | DDC 153.4/3—dc23
LC record available at https://lccn.loc.gov/2021004488

ISBN: 978-0-367-74981-1 (hbk)
ISBN: 978-0-367-74982-8 (pbk)
ISBN: 978-1-003-16061-8 (ebk)

Typeset in Times
by codeMantra

This book is dedicated to my wife, June Linowitz. She shows infinite patience with me as I age. So do my son Oren and daughter Susan. And those married to them – Tara and Joseph. And also my five grandchildren – Jacob, Gabriel, Lee, Zachary and Ben.

They all have helped me with my attempts at problem solving, and discovering new ways to approach problem solving. What could be better than hearing – can I help you with that Grandpa?

Contents

Preface xiii

Author xv

1 The Nature of Problem-Solving 1

Five Large-Scale Problems 2

    Predatory Pricing 2

    Getting Through College 3

    The Climate Issue 3

    Cybersecurity 4

    The Crisis in IT (Information Technology) 4

Problems from the Author's Experience 5

    Problems for the Systems Engineer 8

    Top Department of Defense (DoD) Management Problems 9

Overall Concept of This Treatise 9

Generic Stepwise Approach to Problem-Solving 10

Solving Problems is Our Primary Focus; Thinking Supports 11

Exercises 12

References 12

2 Thinking as a Corporate Culture 13

Thinking as Necessary in Problem-Solving 14

The Learning Organization 14

Moving on to Hewlett-Packard 15

Design Thinking – Change by Design 16

Annual Reports 17

    Xerox 17

    Amazon 18

    Microsoft 18

    SAS (Institute) 19

    Northrop Grumman 19

    Alphabet (Google) 19

    Intel 20

    Lockheed Martin 21

    Apple 21

    General Dynamics 22

Raytheon 22
Leidos 22
Themes 23
Exercises 23
References 24

3 The Power of the Idea **25**
Innovation Revisited 25
Better Idea Stories 26
Netflix 26
IBM: Watson, Sr. and Jr. 26
Xerox 27
Apple 27
Amazon 27
Facebook 28
Google (Alphabet) 28
Oracle 28
SAS 28
FedEx 29
A Group of Generic "Problems" and "Solutions" 29
Blockbuster and Netflix 29
Industry and the DoD 29
What Business Are We In? 30
IBM and Microsoft 30
Dr. Deming and Japan and the US 31
Xerox PARC and Steve Jobs 31
Saul Steinberg and Leasco 31
Coke/Classic Coke, Old Taste 32
Good Idea, Failed Follow-Through 32
Acronym for the Individual with Potentially Better Idea 32
Exercises 33
References 33

4 Approaches to Problem-Solving **35**
The "N step" Disciplines 35
Technical Decomposition/Reductionist 36
Modeling and Simulation 37
Lateral Thinking and (Lateral) Problem-Solving 38
Total Systems Intervention 39
Generalized/Systems Approach 40
Design Approach 41
Expert Systems 41

Definitive Mathematics and Statistics-Based Methods   42
  Statistics-Based Approach   43
DoD-Suggested   43
  How is That for Out-of-the-Box Thinking?   45
Decision Support Systems   45
Cost-Effectiveness Analysis   46
Exercises   47
References   47

5 Think Tanks and Problem-Solving   49
Federally Funded R&D Centers   49
Systems Engineering Research Centers   50
The MITRE Corporation   51
  MITRE Awards over the Years   52
The Aerospace Corporation   52
The RAND Corporation   53
ANSER   54
IDA   55
CNA   56
Energy National Labs   57
Exercises   57
References   57

6 Specific Problems and Their Solutions   59
The Coin Weighing Problem   60
Types of Algorithms   60
The Kalman Filter   61
The Minimum Computer Step Problem   62
The Matchstick Problem   62
The Efficient Message Coding Problem   63
The River-Crossing Problem   63
Parameter Dependency Diagramming   64
  The National Aviation System   64
  A Radar Submodel   65
An Optimal Search   66
Laplace and Fourier Transforms   67
Lagrange Multipliers   67
Allocation of Requirements Errors   68
A Walk in the Park   68
  The Bottom Line   69
Exercises   69
References   69

7 Great Thinkers and Problem-Solvers **71**
Selected Thinkers and Problem-Solvers and How They Thought   71
   Da Vinci   71
   Einstein   72
   Newton   73
   Descartes   74
   Feynman   74
   Edison   74
   Russell   75
   Hawking   75
   Franklin   76
   Jefferson   76
   Socrates   77
   Saint Augustine   78
   Aristotle   78
   Plato   79
   Darwin   79
   The Bottom Line   80
References   80

8 Artificial Intelligence **81**
The DoD Artificial Intelligence Center   82
   The JAIC and AI Technology   83
   Seeing the Forest and the Trees   83
   Overview of Expert Systems   84
   The State of AI in the Enterprise   85
   A Research Agenda   87
   The Future of AI   87
Exercises   90
References   90

9 Group Problem-Solving **91**
GroupThink   91
The DoD and Group Thinking   92
Group Decision Support Systems (Software)   94
   Commercial Decision Support Systems   95
   Collaborative Software Systems   96
Self-organizing and High-Performance Groups   97
Exercises   98
References   99

10 Miscellany **101**
Innovation   101

Innovation in Colleges and Companies 102
Fordham Survey 103
America's Innovation Strategy 104
Problem-Solving Strategies 105
Vision 2030 106
DARPA 106
The Eight Disciplines 108
The Lemelson Center 109
Military Design Thinking 109
Appreciative Inquiry 109
Grounded Theory 110
Morphological Analysis 111
The 2×2 Matrix 111
Synectics 112
Problem-Solving and Re-engineering 113
Cost Effectiveness 114
Cost Assessment Data Enterprise 115
Measures of Effectiveness 115
MOEs for a Communications System 115
MOEs for a Transportation System 116
Leontief Model 116
Commentary on Systems of Systems 116
Problem-Solving by Reduced Clock Speed 117
George Washington University Class Using "Thinking" as Textbook 118
Courses Meet via Blackboard Collaborate Ultra 118
Exercises 119
References 119

11 Summary **121**
Twelve Problem-Solving Approaches 121
Twelve Thinking Approaches 122
Another Twelve Thinking Approaches 122
Exercises 123
References 123

Appendix A: A Dozen Additional Ways of Thinking 125
Appendix B: Acronyms 141
Index 143

Preface

Problem-solving skills are among the most sought after by today's employer. Yet, it is difficult to predict who will do it well, who poorly and who not at all.

What are some scenarios we can explore to dig a bit deeper? Let us look briefly at possible examples from the world of work, home and school.

At work, consider the following: (a) your next assignment, as part of our strategic plan, is to do the SWOT (strengths, weaknesses, opportunities and threats) analysis; (b) your boss has expressed unhappiness with your marketing plan and results; (c) in your line of business (LOB), your revenues are down 12 percent for the last quarter. Are you prepared to tackle all these problems?

At home, you are faced with the following: (a) your furnace appears to be not heating or cooling; (b) you need a fire escape solution, including a permit that is not forthcoming; (c) you wake up sneezing every other day. So what is the problem? You just have to call a person to come and solve. The problem is that the virus has led to your job loss, and you are almost out of money. What to do?

At school, you are confronted with (a) a term paper for which you have not a word on paper, and it is due tomorrow; (b) your study team has lost a critical member; (c) your professor wants to talk to you as soon as possible about your last assignment, which you were not comfortable with. Do you have a problem? So far, not known.

In this treatise, we look at problem-solving and these kinds of interconnections:

- In relation, primarily, to thinking
- In relation to creativity
- In relation to situation and context
- In relation to re-engineering
- In relation to experience

And we provide a bottom line, top dozen list and discussion – what are the twelve problem-solving approaches that appear to lead to the greatest likelihood of success? We hope the reader's attention is captured, and that the read is an interesting and useful adventure.

Howard Eisner
Bethesda, Maryland

Author

Howard Eisner spent 30 years in industry and 24 years in academia. In the former he was a working engineer, manager, executive (ORI, Inc., and the Atlantic Research Corporation) and president of two high-tech firms (Intercon Systems Corporation and the Atlantic Research Services Corporation). In academia he served as professor of engineering management and distinguished research professor in the engineering school of the George Washington University (GWU). At GWU, he taught courses in systems engineering, technical enterprises, project management, modulation and noise, and information theory.

He has written ten books that relate to engineering, systems and management. He has also given lectures, tutorials and presentations to professional societies such as INCOSE (International Council on Systems Engineering), government agencies (such as the Departments of Defense and Transportation, and NASA), and the Osher Lifelong Learning Institute (OLLI).

In 1994 he was given the outstanding achievement award from the GWU Engineering Alumni.

Dr. Eisner is a life fellow of the IEEE (Institute of Electrical and Electronics) and a fellow of INCOSE and the New York Academy of Sciences. He is also a member of Tau Beta Pi, Eta Kappa Nu, Sigma Xi and Omega Rho, various honor/research societies. He received a Bachelor's degree (BEE) from the City College of New York (1957), an MS degree in electrical engineering from Columbia University (1958) and a Doctor of Science (DSc) degree from the GWU (1966).

Since 2013, he has served as professor emeritus of engineering management and distinguished research professor at the GWU. As such, he has continued to explore advanced topics and write about engineering, systems and management.

The Nature of Problem-Solving

1

Problem-solving is one of the most important skills, say the employers of today. It stands out among other skills, which are attributes like [1]:

- Positive attitude
- Communication
- Teamwork
- Self-management
- Willingness to learn
- Resilience

Now that is quite a list; fortunately, we will deal only with problem-solving in this book. But that appears to be sufficient.

And so we find that Forbes declares [2]:

"Problem solving is the essence of what leaders exist to do...their goal is to minimize the occurrence of problems". Yes, one solution is to keep them from happening.

We will feature a dozen ways to deal with problem-solving, each one requiring considerable effort to use and master. In addition, these techniques are enhanced if the user brings improved thinking to the fray. This book identifies some two dozen potentially new ways of thinking. A lot of work for the reader. The author hopes that all these efforts will prove to be useful with a tangible payoff. On the matter of effort, we take note of the apparent fact that effort is a "multiplier" [3]:

"talent × effort = skill"   and   "skill × effort = achievement"

We note that Angela Duckworth put forth the proposition that achievement is just about proportional to the square of the effort. The math details perhaps are not important, but the basic idea is. You have got to put in the effort if you want to achieve.

So if you are the boss in an enterprise, you would like your direct reports to have this skill, in spades. Your boss drops an important and complicated

1

problem on your desk and you would like to get some help in responding. You go to your people and drop some aspect of the problem on their desks. Two or more heads are better than one, and they are. So your guys work over the weekend to get a handle on the problem. By Monday afternoon you are presenting some preliminaries to your boss. You are pleased, and so is she. The "machine" is working the way it should. The "machine" is working as if it was well-oiled, and it is. All is well in your territory. And it is based upon orderly thinking and problem-solving. Without these skills, the machine is lying on the ground, huffing and puffing. With the machine in tune, you are happy and so is your boss.

FIVE LARGE-SCALE PROBLEMS

Two types of problem areas are explored in these two sections. The first has to deal with large-scale issues that we have as a society. Each of us as individuals experiences the problem, but mainly as a member of society at large. The second problem area is a set of problems that this author faced during his 30 years in industry. This latter set provides a series of examples that are illustrative. The first set has the following elements:

1. Predatory pricing
2. Getting through college
3. The climate issue
4. Cybersecurity
5. The crisis in IT (Information Technology)

Predatory Pricing

Various small businesses, especially, have run into what they have called "predatory pricing" [4]. They experience the phenomenon directly, they claim, when they have a best-selling product and, all of a sudden, a large competitor appears and undercuts their price by 10%–15%. Same product, basically, but cheaper by 10%–15%. We know what the consumer is going to do. He or she is going to switch to the cheaper offering. Then, the producer gets in touch with the complainer and offers to make a deal whereby they "share" the market. The small business now has a double problem – make a deal or do not make a deal? Complain to the government, or do not complain to the government?

Do a bunch of analyses on the problem, or do not spend more time with numbers and projections? So the basic problem that the small business faces has a variety of subproblems, or questions. The small business has a problem that, if handled poorly, could determine whether or not they succeed as a business.

Getting Through College

In case nobody noticed it, we remind everyone that college prices have been steadily increasing. Many students now expect to have a large debt facing them after they graduate from college. How to "solve" such a problem? It is not obvious, but here are some "solutions": (a) skip college for now until the value-proposition is more favorable; (b) borrow the money from your parents, or other source; (c) go to a less expensive 4-year college; (d) go to a "community" college, which normally is less expensive and plan to move to your college of original choice, based upon your terrific grades and a scholarship. There are probably other variations on the theme, if you think hard enough. So what is the answer? It would appear that just about each and every student has his or her own answer. And so we struggle along.

We continue with a short citation of numbers relative to this problem area [5]. During the past 20 years (*):

- The average tuition and fees at private national Universities have jumped 154%
- The average tuition and fees at public Universities have risen 181%
- In-state tuition and fees at public national Universities have grown the most, increasing 221%
- (*) sample size of 381 Universities

The Climate Issue

One more time, the new report rings the bell on the "climate" problem. One report identifies ways that climate change can affect the ocean [6]. One way, of course, will lead to unacceptable new shorelines. Others will just about kill certain areas for productive fishing. Yet a third will lead to horrific melting of polar ice.

Negative consequences are right around the corner, such as:

- Increasing sea levels and atmospheric water vapor
- Decreasing mountain glacier and arctic ice mass, snow cover and permafrost

- Extreme weather such as droughts, heat waves, wild fires, storms, floods, and blizzards
- Aggregate negative effects on human quality of life

The research-oriented crew at Sigma Xi have decided to increase their support for such actions as [7]:

- Climate scientists as they produce their results
- Global cooperation and collaboration
- Policies and investments that enable technological solutions

Cybersecurity

With the increase in the world of IT (information technology) every company now has to worry about cybersecurity. Hackers seem to be everywhere and it comes down even to the individual and stories that we hear about stealing of identity, breaking into bank accounts and making off with passwords and checks from the local mailbox.

A very recent example of what we are doing about this problem is the new comprehensive policy called a Space Directive – 5 to improve the nation's defense against cyberattacks [8]. This is focused on space and near-space technologies, and it puts the Department of Homeland Security clearly in charge of cybersecurity defenses. As part of the new Directive, key cybersecurity principles were established, to include:

a. Space systems are to be subject to risk-based cybersecurity-informed engineering
b. Space systems operators must develop and follow cybersecurity plans
c. Space systems need to be subject to best practices as well as possible new regulations
d. Space systems owners and operators will be required to manage to well-defined risk tolerances and minimize undue burden on users of various types

The Crisis in IT (Information Technology)

The September 2020 issue of the *IEEE Spectrum* has a lead article that is called IT's perpetual crisis [9]. It refers to the huge amounts of hardware and software that are no longer supported by the originators or by maintenance folks. As the article points out, these systems are subject to failures, outages and errors, which translate into problem for the ultimate users, the general public. That means banks, insurance companies, investment houses and all manner of services. Some numbers that are quoted: since 2010 corporations and governments worldwide have spent an estimated $35 trillion on IT products and services. About three-quarters of this went toward O&M (operations and maintenance) existing systems. Some $2.5 trillion was used to try to replace legacy systems, of which some $720 billion was wasted on failed replacement activities. Is this problem likely to go away with time? The assessment is that it is likely to become worse as more systems are needed and as the economy expands.

PROBLEMS FROM THE AUTHOR'S EXPERIENCE

To further illustrate the variety of problems that have called for "problem-solving" skills, we briefly cite some of the contracts that the author has worked on over the years. Ten of them are listed below. All of them required skills that one would label "problem-solving" skills along with appropriate methodologies.

1. The Nimbus meteorological satellite
2. The Interrogation, Recording and Location Satellite (IRLS) System
3. The Federal Aviation Administration (FAA)'s air traffic control (ATC) radars
4. A Model of the National Aviation System (NAS)
5. Mallard, a battlefield communications system
6. The Climatic Impact Assessment Program (CIAP)
7. The Aviation Advisory Commission's (AAC) Study
8. The Joint National Aeronautics and Space Administration (NASA)-Department of Transportation (DOT) Intercity Transportation Study
9. The Performance of the Basic Point Defense Surface Missile System
10. A Generic Model of the Strategic Defense Initiative (SDI) Program

The Nimbus Satellite. Nimbus was a three-axis stabilized weather satellite that came after TIROS, a spinner. We were asked to do a reliability assessment of the satellite. Generally, that led to a couple of approaches:

- Estimate the reliability of each subsystem and the entire satellite
- Via detailed design reviews, see where changes, such as single-point failure designs and use of nonspace qualified components, would be beneficial
- Determine whether or not the spacecraft and launch vehicle are likely to collide after some number of orbits

The IRLS System. This was a position location satellite. Some of our activities included:

- A design review of the system
- Providing algorithms for the position location function
- Performing an error analysis in order to understand the error sources as well as how to estimate the errors in position location

FAA Air Traffic Control Radars. These were the FAA Airport Surveillance Radar (ASR) Radars. The primary question was how to monitor the performance of these radars. Might any changes be made that would improve upon their current ways of doing the monitoring. The result was "yes" and led to a Radar Quality Control Feasibility Experiment. That experiment involved such activities as:

- Collection of real world radar data for 60 days of operation
- Analyzing the data collected
- Drawing conclusions about the performance of these radars
- Making appropriate recommendations, based upon what the data were suggesting, to include constant monitoring

NAS Model. The FAA wished to have such an overview model built. The author's company won the competition and built the preliminary model, with the following f functionalities:

a. airport capacity
b. airspace capacity
c. delay
d. ATC availability
e. trip time
f. energy utilization
g. service availability
h. noise
i. air pollution
j. security
k. safety

l. demand
m. cost

We note that each of above required a submodel and ways to combine them for special issue studies.

The Mallard Battlefield Communications System. The author's firm was a subcontractor on a team headed by GT&E, and with IBM. The basic problem was to design the next generation of battlefield communications system, top to bottom. This included the design of all the major functions such as those listed below.

a. User access
b. Modulation/demodulation
c. Multiplex/de-multiplex
d. Transmission
e. Encryption/de-encryption
f. Networking
g. Telephony
h. System monitoring and control

It also involved building a life cycle cost model so that the costs of this future system could be properly assessed.

The CIAP. The overall problem was to be able to compute the effects on the climate of a fleet of supersonic aircraft. The approach was basically canonical – break the problem into 4–5 subproblems of conditional probabilities. The first element was building a scenario of SuperSonic Transport (SSTs), each leaving effluent deposits in the stratosphere. The second part was the effects on the air chemistry from those effluents. The third part was the transport of these effects to people on the ground. And the third part was those influences, quantitatively, on the degree of cancer with respect to those people. This study was run out of the Office of the Secretary in the Department of Transportation and involved the participation of many research institutions across the country.

AAC Study. The charter of the AAC was to look at the possible future of our National Aviation system and try to move it in a better more efficient direction. We assisted the commissioners by helping them with such problem areas as:

a. What are the reasonable new future scenarios for the national aviation system?
b. Will airports play the same exact role, or will new types of airports come into being?

c. Can we conquer the noise and air pollution problems, or at least make improvement in these domains?

d. Are new types and mixes of aircraft called for in the system of the future? For example, V/TOL, STOL, helicopters, etc.

e. Do we need new institutions to move toward the system of the future?

We can see a whole host of issues in the above, where we go beyond purely technical matters and into social and economic considerations.

The Joint NASA-DOT Intercity Transportation Study. In this activity, NASA and the DOT teamed up to take a new look at problems and issues of intercity transportation. The author won a contract to be a participant in this investigation. The basic problem addressed here was:

• What is the best way to carry out intercity freight transportation?

The essential method to find an answer was to compare five freight systems. New measures of effectiveness were developed for this comparison. It took about 2 years to execute this investigation.

The Basic Point Defense Surface Missile System. The Captain in charge wanted us (the author's company) to develop handbooks for the system that would help his engineers maintain and troubleshoot the system. So we built a "model" of the system. It explained various aspects of system operation. It was a quite challenging set of problems to "solve".

The SDI Program. President Reagan had just set this program in motion. We started to build a model of the overall scenario. We were visited by a large company that appeared to be very interested in our model and its conception. They engaged us as a subcontractor. Only good things happened after that, as we were well on our way to solving a very difficult problem.

Problems for the Systems Engineer

Faster, Cheaper, Better. So Dan Goldin came to town as the new NASA administrator and declared that it was his charge to build systems faster, cheaper, and better. That meant with shorter schedules, below cost, and with stronger performance profiles. That is a problem for today's systems engineer while some believe that two of the three are feasible, but three of the three are not.

Better Systems Architecting. The challenge here is that we need to improve radically upon how we do systems architecting. This very likely means that we may have to give up on the Department of Defense Architecture Framework (DoDAF) model and move to an alternative.

Technology and Risk. We need greater insight as to how to do the tradeoff between technology and risk for our systems. Is there a way to do this quantitatively?

Design for Integration. Since integration is such a serious problem, we need to design our systems so that they are more easily integrable. How do we do that?

Cyber-Security. Given todays' trends, all systems have to be built for high cybersecurity. What are the preferred configurations for the super-secure systems. Note Cyber-Directive above.

High-Performance Teams. We have been talking about these entities for a while. How do we build and maintain team as standard across the industry?

The "ilities". We thought there were enough "ilties" to worry about. Are there some new ones that join our list. How about sustainability, affordability, and others?

Top Department of Defense (DoD) Management Problems

The DoD called them management challenges [10] but this author has converted them into problems. The Office of Inspector General has them enumerated:

- Separating state actors and terrorism
- Space- and ground-based missiles with nuclear capabilities
- Balancing force structure, modernization, and readiness
- Transition from efficiencies to reform initiatives
- Financial management
- The continuing problems of cyber, contracting, healthcare, and ethics

The above sections indicate that there appear to be enough problems for the problem solvers to grabble with, or in other words, problem solving is an important skill to have.

OVERALL CONCEPT OF THIS TREATISE

This being the introductory chapter, we pause for a moment to sketch the overall concept of this book. Basically, the centerpiece is the set of some twelve approaches to problem solving, as shown in Figure 1.1. This is supported by

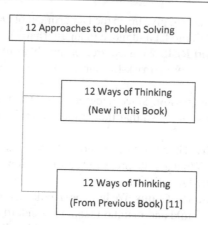

FIGURE 1.1 Approaches to problem-solving and ways of thinking.

two sets of twelve ways of thinking, also depicted in Figure 1.1. All of this is reiterated in Chapter 11, Summary. In that case, the specific approaches and ways are enumerated.

GENERIC STEPWISE APPROACH TO PROBLEM-SOLVING

Several approaches out there in the real world put forth what might be called the "step-by-step" approach. Not surprisingly, these approaches are quite similar. We identify them with the following "generic stepwise" approach and recognize that other approaches are likely to have more steps and be more complicated. In fact, we will see these two constructs with such approaches like the "8 disciplines" and the DoD recommendations that come in later chapters.

Step One – STATE THE PROBLEM
Step Two – ESTABLISH KEY VARIABLES AND DEPENDENCIES
Step Three – IDENTIFY ALTERNATIVE SOLUTIONS/ APPROACHES
Step Four – SELECT AND IMPLEMENT BEST ALTERNATIVE
Step Five – DOCUMENT AND PRESENT RESULTS
 State Problem. This needs to be done succinctly and with clarity. Too many people have this as fuzzy, and of course, they wind

up working on the wrong problem. What are the chances of being successful?

Establish Key Variables and Dependencies. Key variables are outputs (most important), inputs, and intermediate parameters. Then we want to know as much as possible about the nature of relationships between all of them.

Identify Alternative Solutions. You cannot get there without realizing the existence of alternatives. Be as clear as possible in this regard. Be aware of other approaches (like the AoA – Analysis of Alternatives - from the DoD).

Select and Implement Best Alternative. Focus on what appears, at the beginning, to be the best approach/answer. You may guess incorrectly, but you must "bite the bullet" in this regard. If you guessed wrong, you may have to go back and re-trace your steps. Possibly you will find that another alternative should have been selected and implemented.

Document and Present. Once you have your answer, you need to have good documentation (and presentation) skills. They are both invaluable in getting to the finish and credibility line.

SOLVING PROBLEMS IS OUR PRIMARY FOCUS; THINKING SUPPORTS

This chapter attempts to make the point that the problems we can face are likely to range from difficult to very difficult; from challenging to extremely challenging. Help in problem-solving would be most welcome. The author hopes we are up to that challenge.

The overall process we are suggesting in terms of problem-solving is to:

a. First select a problem-solving approach from among the twelve set forth in Chapter 4
b. Then select a thinking approach from Chapter 4
c. Then select another possible thinking approach, also from Chapter 4

The first thinking approach is that covered in the author's book [11] on thinking. The second thinking approach is new. The bottom line is that you are possibly tempering your problem-solving approach with one or more thinking approaches.

Here are the next chapters leading up to the last chapter, number eleven, which is the Summary. That is followed by two Appendices.

Chapter Two – Thinking As a Corporate Culture
Chapter Three – The Power of the Idea
Chapter Four – A Dozen Approaches to Problem-Solving
Chapter Five – Think Tanks and Problem-Solving
Chapter Six – Thinking: Let us Solve
Chapter Seven – Great Problem-Solvers
Chapter Eight – Artificial Intelligence
Chapter Nine – Group Problem-Solving
Chapter Ten – Miscellany

EXERCISES

1. Identify and discuss, in three pages, problems that you feel that you need to address with respect to your everyday life
2. Do the above with respect to national problems
3. Of the various N-step problem-solving processes, which ones do you find most attractive? Why? Document in three pages.

REFERENCES

1. See www.careers.gov.sg; "Skills Employers are Looking For".
2. Llopis, G., "The Four Most Effective Ways Leaders Solve Problems", Forbes, Nov. 4, 2013.
3. Duckworth, A., *Grit*, Scribner, New York, 2016.
4. See https://truthonthemarket.com; Stout, K. and A. Stapp, "Is Amazon Guilty of Predatory Pricing?"
5. Boyington, B., and E. Kerr, "20 Years of Tuition Growth at National Universities", see www.usnews.com.
6. Pink, J., "5 Ways the Climate Change Affects the Ocean", see www.conservation.org.
7. See www.sigmaxi.org; "Science Statement on Climate Change", Sigma Xi.
8. See public.cyber.mil/dodd-3.
9. Charette, R., "IT's Perpetual Crisis", IEEE Spectrum, September 2020.
10. See https://www.dodig.mil/Reports/Top-DoD-Management-Challenges/Article/2108369/top-dod-management-challenges-fiscal-year-2020/.
11. Eisner, H., *Thinking – A Guide to Systems Engineering Problem Solving*, CRC Press, Boca Raton, FL, 2019.

Thinking as a Corporate Culture

2

Thinking as a matter of corporate culture is a form of group think but is distinctly not the kind that is the same as GroupThink. This is positive thinking as a matter of the culture of an organization. This refers to a company's attribute where everyone is encouraged to do their best thinking as a part of the job, twenty-four by seven. We begin by looking at the champion of cultural thinking in a company, namely, Thomas Watson, Sr. Leaving National Cash Register and moved on to IBM in search of an enterprise in which he could imprint a culture. That turned out to be IBM and the culture was that of to "think". To think was the cornerstone of the basic beliefs of the company, namely:

- Excellence in all products (and related services)
- The best possible customer service
- Respect for the individual, both within and outside the company

Watson Sr. eventually gave way to his son, Watson Jr., who had an even more illustrious career as the president of IBM (in 1952) and other activities that were indicators of the man's range of interests as well as competence. Watson Jr. brought IBM from the day of punched card systems to the computer. He led the charge in that field through contracts with the Air Force and its SAGE tracking system. It was the leader in the field, bringing the company from revenue levels of 214.9 million in 1950 to 734.3 million in 1956. Although the computer was the most dramatic product, it was not the only one for which IBM was noted. One remembers the Selectric typewriter as well as its tendency to invest in high-tech devices through an aggressive invention and patent program. They hit some roadblocks as they entered the 1990s and the era of Louis Gerstner. He "turned the company around", using what he had learned earlier at a biscuit company. How is that for lessons learned? He was indeed credited with teaching the "elephant to dance".

13

IBM remains a very strong hi-tech company and many attribute its longevity to its basic culture (think) and its fundamental beliefs. This author is among its believers and agrees that it is possible and desirable to work on the matter of establishing a coherent culture for just about any enterprise.

Watson Sr. set the stage for dealing with IBM's culture and hammered away at it any chance he could. It has been said that his employees were often reminded – think please, think! Then "think" became a company trademark. Then came the ThinkPads. Beyond that, a line of desktop computers became a ThinkCentre. So it is one thing to decide that "thinking" should be a definitive part of your culture. It is yet another to make it happen. Do you want your organization to strive for a "thinking" culture and work on making it happen?

THINKING AS NECESSARY IN PROBLEM-SOLVING

The ultimate capability, many claim, is having an organization that is able to solve problems. That is what it is all about, they assert, since no enterprise runs smoothly all the time. Problems are inevitable, but solutions are not. So the position here is that we are on our way to problem-solving, but on that journey, we have a critical stop-over – deal with establishing a thinking enterprise. This has many dimensions, as the following chapters demonstrate.

THE LEARNING ORGANIZATION

If "think" is the key element of a company's culture according to Watson, then the "learning organization" is the key cultural element according to Peter Senge [1]. It is what an enterprise should desire to become; it should be the answer to the questions:

a. Who are we?
b. Who do we wish to become?

Senge implicitly accepts the notion that becoming a learning organization is not only a good thing; it is a critical thing. It is a thing not to be questioned.

It is a thing to strive for – hence all work on what it is and how to achieve it. As to what it is, Senge goes on to say that we may achieve the learning organization by means of five disciplines, as articulated in the previous chapter. The fifth discipline is "systems thinking" which we will explore more deeply in Chapter 4. The bottom line, for now, is simply that a learning organization is an irrefutable goal and an important aspect of achieving this state of affairs is to bring "thinking" into the corporate culture in a most serious way.

MOVING ON TO HEWLETT-PACKARD

Hewlett-Packard is an outstanding company with some ups and downs, but mostly ups. Dave Packard has done us all a great service by writing the treatise "The HP Way" [2] that takes the reader to the 1990s from literally the beginning in the garage in Palo Alto along with being students at Stanford and being encouraged to start a company by the then Dean, Professor Terman.

Packard's overview of HP can be summarized by the following points:

a. Management by Objectives
b. An open-door policy, including management by walking around
c. Participative management
d. Emphasis on innovation in terms of a stream of new products
e. Listening to the customer
f. Trusting its employees
g. Internal cooperation
h. Recognizing individual achievements
i. Continuing education and self-development

Quite a list, especially when one considers how well Hewlett and Packard maintained forward motion and year-by-year consistency.

What is the corporate culture that one can glean from the above list? The label I would use is simply respect for the individual, both employee and customer. If we add very strong hiring practices to the mix, we wind up with all the ingredients needed to produce a leader in today's high-tech industry. And the emphasis on innovation assured that they were "thinking" all the time. And Packard's book does not neglect another important aspect of the company – responsibility to society.

DESIGN THINKING – CHANGE BY DESIGN

If we wish to see a company dominated by "thinking" we need not go any further than the "Change by Design" enterprise called IDEO [3]. This firm is based upon the notion that products (and services) can be re-designed for the better, and such that customers are indeed willing to pay substantial sums for the new design. Coming up with an exciting new design takes a lot of thinking, and often on the parts of many contributors. Some of the notions behind this company have to do with:

a. Many people on the design team, working individually and also as a group
b. Definition of many alternatives
c. Use of sketches and drawings
d. Storytelling
e. Human-centered
f. Use of prototyping

So we can imagine that a new product question is put in front of a team of some dozen people and all members of that team react immediately, and also for or against other ideas form other people. The leader keeps it from becoming a free-for-all, but if it does, a recess is called, possibly for a whole day. Everyone is constantly thinking and contributing in a wide circle that eventually narrows and converges. There is a fair amount of data that are collected during the process, for example, about similar projects and products as well as demand information.

To illustrate the process, imagine that we are re-designing the ubiquitous grocery cart, a favorite activity from time to time. We are given some ground rules that suggest that the current design will not be exactly right. So we go through the "change by design" process and come up with a new design with the following features:

a. It becomes a consumer product instead of grocery store owned
b. Much smaller than current baskets (about half the size)
c. Two wheels, each with $360°$ swivel
d. Straps to attach unit to waist of shopper while shopping
e. Folds up to lie flat
f. Especially useful to bring groceries from car trunk to kitchen

So we see that there is no end to the imagination when we are dealing with food and its transport.

ANNUAL REPORTS

For the remainder of this chapter, we pause to look at aspects of the annual reports of several high-tech companies. The notion is that companies reveal who they are and what they consider important through the messages in their annual reports. Can the reader see common themes in these reports?

Xerox

"Our three year plan guided by our four strategic initiatives which are to:
optimize operations, drive revenue, re-energize our innovation engine, focus on cash"
"built a leadership team"
We help our customers to "make the right choices on workplace technology that makes them more efficient, cost-effective and more secure"
"We provide our clients with innovative solutions that combine our well-established print technologies with a growing portfolio of software and services"
"We also invest in new lines of products to offer clients better productivity and image quality at a lower total cost of ownership"
"We will continue to introduce new innovations and enhancements to expand market share"
"We are re-energizing our innovation engine with a focus on five key areas":

1. 3D printing and digital manufacturing
2. AI workflow assistants
3. IoT Sensors and services
4. Digital packaging and print
5. Clean technology

"We celebrate PARCs fiftieth anniversary"
"We create value for business and society"
"We support Fortune 500 companies that want to improve sustainability, hospitals that need to digitize paper health records and clean technologies that have a profound effect on reducing greenhouse gasses"
"We continue our strong pipeline of M&A (mergers and acquisitions) activity"

Amazon

"We are working around the clock to get necessary supplies delivered directly to people who need them"

"We quickly prioritized the stocking and delivery of essential household staples, medical supplies and other critical products"

"Our Whole Foods Market Stores have remained open...but we have temporarily closed Amazon Books, Amazon 4-star, and Amazon pop-up stores because they don't sell essential products"

"We are focused on the safety of our employees and our contractors"

"We've made over 150 significant process changes in our operations network and Whole Foods Market stores to help teams stay healthy"

"We've begun the work of building incremental testing capacity"

"We opened 100,000 2011 positions across our fulfillment and delivery network"

"We are collaborating with the World Health Organization and recruiting other companies to sign The Climate Pledge"

"We aim to have 10,000 of Rivian's new electric vans on the road as early as 2022 and all 100,000 vehicles on the road by 2030"

"We're committed to reaching 80% renewable energy by 2024 and 100% by 2030"

"We are making significant investments to drive our carbon footprint to zero"

"Over the last decade, no company has created more jobs than Amazon. We employ 840,000 workers worldwide"

"We raised our minimum wage to $15 per hour, an industry leader"

Microsoft

"Our mission is to empower every person and every organization on the planet to achieve more"

"We work alongside our customers and partners to help them build their own digital capability"

"The era of the intelligent cloud and intelligent edge is shaping the next phase of innovation, powering intelligent systems"

"We are innovating to empower our customers"

"The variety, velocity and volume of data is increasing – with 50 billion connected devices coming online by 2030"

"The quintessential characteristic for every application going forward will be AI"

"Microsoft 365 empowers everyone, with an integrated secure experience that transcends any one device"

"We are pursuing our expansive opportunity to transform how games are distributed, played and viewed"

"Trust begins with our commitment to shared success and prosperity"

"The first pillar is privacy. The second pillar is cybersecurity. And finally, we build AI responsibly"

"We continue to operate carbon neutral across our worldwide operation"

"At its core, Microsoft's strength lies in its talented people"

SAS (Institute)

"We're partnering with organizations worldwide to put analytics to work where it's needed most"

1. In hospitals
2. With epidemiologists
3. With government
4. With agencies using contact tracing
5. With health systems
6. Producers of vital goods like food and medical supplies

Northrop Grumman

"We successfully integrated Orbital ATK in its first full year of operations"

"We focused on areas such as space, cyber, strategic missiles, and all-domain command and control"

"This new structure is designed to accelerate our ability to rapidly identify and deliver the technologies, products and services that our customers need"

"Our new sectors are Aeronautics systems, Defense Systems, Mission systems, and Space systems"

"Defining what is possible is our purpose, and discovery and innovation are at the heart of everything we do"

"We continue to build on our strong culture with leading environmental, social and governance practices. This culture is reflective of who we are"

Alphabet (Google)

"Alphabet is a collection of businesses the largest of which is Google"

"Far afield from our main internet products in areas such as self-driving cars, life sciences, internet access and TV services"

"We're making sure our core Google products are fast and useful"

"We acquired YouTube and Android, and when we launched Chrome... those efforts have matured"

"We will not shy away from high-risk high reward projects"

"Machine learning and artificial intelligence are increasingly driving many of our latest innovations"

"We're committed to building products that have the potential to improve the lives of millions of people"

"Google's core products and platforms such as Android, Chrome, Gmail, Google Drive, Google Maps, Google Play, Search and YouTube each have over one billion monthly active users"

"Our vision is to remain a place of incredible creativity and innovation"

Intel

"We are at a strategic inflection point where the emergence of data is a transformational force"

"This includes the iconic Intel culture that helped define not only our company but Silicon Valley itself. Our culture is more differentiating than ever, particularly the value we place upon brilliant engineering, integrity, truth, and transparency"

"We must approach each day with a growth mindset"

"Our unique ability to reimagine the boundaries of innovation, achieve economies of scale and build global ecosystems"

"Play a much larger role in our customer's success"

"Significant investments in R & D"

"That's why we're infusing AI into everything we build"

"Execution, inspired by Moore's Law" (he was one of the founders of Intel)

"New products, platform initiatives, and software optimizations":

1. Data center performance
2. Client performance
3. Software optimization

"evolving our culture":

1. Must be customer obsessed
2. Must show up as "one Intel"

3. Must be fearless
4. Needs truth and transparency
5. Must create diverse and inclusive workplace

Lockheed Martin

"We put our customers at the center of everything we do"

"Our ongoing research and dialogue helped reinforce the critical role or our products, programs, and capabilities in protecting citizens, creating jobs, and driving economic growth"

"At Aeronautics, we are delivering the most advanced aircraft and aviation technologies in the world. Nowhere is this leadership see more clearly than in the growth and impact of our globe-spanning F-35 program"

"The iconic F-16 fighter continues to find new customers"

"We also marked the 50thyear of service for the C-5"

"We won a $1.8 billion contract to upgrade the missile defense capabilities of U.S. and allied military forces"

"We view innovation as the lifeblood of our corporation, with new programs in hypersonics, laser weapons, autonomy and artificial intelligence (AI)"

"We have a commitment to uphold our core values to do what's right, respect others, and perform with excellence"

Apple

"The company is committed to bringing the best user experience through its innovative hardware, software and services"

"We provide products and solutions with innovative design, superior ease-of-use and seamless integration"

"We support a community for the development of third-party software and hardware"

"We believe that ongoing investment in R&D, marketing and advertising is critical to the development and sale of innovative products, services and technologies"

"Key products include":

a. iPhone
b. iCloud
c. AppleCare
d. iPad
e. Mac
f. Apple TV

General Dynamics

"We are committed to invest over the long haul to satisfy the needs of our customers"

"The FAA approved the G600 type and operating certifications"

"We continue to invest in our Aerospace service network and expand the use of sustainable aviation fuel"

"Combat systems continues to deliver systems that meet the top priorities of the U.S. and our allies"

"General Dynamics Information Technology continued to integrate CSRA and to reshape its portfolio"

"Our Mission Systems business stayed at the leading edge of radio frequency and network communications, cyberwarfare and maritime systems"

The Navy awarded its largest shipbuilding contract in history of $22.2 billion to Electric Boat"

"Over the past 5 years, we have expended more than $2.3 billion on independent R&D to bring new products to market and concluded several acquisitions"

Raytheon

"We're pushing the boundaries of technology, working at the forefront of quantum physics, artificial intelligence and machine learning, hypersonics, cybersecurity, and much more"

"We focus resources on emerging opportunities within the DoD market"

"We emphasize capabilities in high energy lasers, high power microwaves, space, hypersonics and next-generation sensors and cybersecurity"

"We focus on creating cost-effective best value solutions"

"We embrace a culture of continuous innovation"

"We are broadening our international marketplace"

"We have a legacy of innovation, including laser radars, one of more than 13,000 active Raytheon patents"

Leidos

"We focus on rapidly deploying agile, cost-effective solutions to meet the ever-changing missions of our customers in the areas of intelligence, surveillance and reconnaissance, enterprise information technology, integrated systems, cybersecurity and global services"

"We deliver innovative technology, large-scale intelligence systems, command and control platforms, logistics, and cybersecurity solutions"
"We concentrate on":

1. Digital information and integrated systems
2. Aviation solutions
3. Security products
4. Technical capabilities
5. Health systems

"We compete on several factors, including technical expertise and security-cleared personnel".

THEMES

We conclude this chapter by scanning the annual reports to cite a couple of themes that are common to almost all of the selected companies.

First on the list of common messages is that these companies are strongly innovative. Just about all of them stress that word and that approach. Second, these companies want to make sure that readers understand that they (the companies) just about always help their customers with the latter's problems. So it is more than just passing off a piece of hardware or software, it is a matter of then asking the question – and now, how can I help you with the problem that you have and the company that you are constantly building?

And on the innovation side of things, it is IBM that tries to prove their innovative culture by pointing to patents, specifically. From an annual report, IBM cites the fact that IBM inventors received 9,262 U.S. patents in 2019 – the most ever awarded to an U.S. company. It was their twenty-seventh straight year of patent leadership. They also indicate that that they work side-by-side with clients to apply IBM Design Thinking and agile methods to finding solutions. Looks like IBM is still devoted to thinking, a legacy that goes all the way back to the Watsons.

EXERCISES

1. Obtain quotes from the annual reports of 6 high-tech companies (other than those in this text) and compare them in three pages of your own text.

2. What common themes do you see in these annual reports? Comment on four such themes in three pages.
3. Which four themes are most important? Why? Answer in three pages.

REFERENCES

1. Senge, P., *The Fifth Discipline – The Art and Practice of the Learning Organization*, *Doubleday*, New York, 1990.
2. Packard, D., *The HP Way*, HarperCollins, New York, 1995.
3. Brown, T., and B. Katz, *Change by Design*, Harper Business, Manhattan, NY, 2009

The Power of the Idea

3

So many things in life depend upon the better, seminal idea. And that, in turn, depends upon how a person or an organization is able to think. In this author's previous book on thinking, a full chapter is devoted to innovation. This book, in the previous chapter, demonstrates how much attention is paid to innovation in the annual reports of several significant companies. That is, these companies are uniformly willing to declare how important innovation is to them. And there is little doubt that this is true. Innovation is the name of the game, and innovation will be the cause of many successes and failures in the years to come. It will not be abandoned as the primary source of success.

INNOVATION REVISITED

We will start by a quick scan into innovation by looking at what Isaacson had to say in his special book on the subject [1]. He focused on the people that he called the "innovators", which is entirely appropriate for our purposes. The list below is a summary of some his innovators and what they achieved, above and beyond the citations in the author's prior book (page 55).

- Claude Shannon – for his work on Boolean logic, switches, and his invention of the field of information theory
- John Mauchly – for his contributions to the early instantiation of the digital computer
- For the team that developed the Transistor at Bell Labs
- For Paul Baran who developed important concepts of packet switching
- For DARPA and its contributions to the Internet
- For Vint Cerf and Bob Kahn and their work on Protocols for the internet
- For Linus Torvalds who constructed the first kernel for Linux

- For Marc Andreessen who built the Mosaic Browser
- For Paul Allen and Bill Gates who built BASIC for the Altair machine as well as sustained growth in Microsoft
- For Ray Tomlinson who invented e-mail
- For IBM, where the Watson computer first beats Kasparov at Chess, and later the Jeopardy champions
- For Steve Jobs, who deserves a second citation for building insanely wonderful products such as the Mac, iPhone, iPad, iPod and Watch

BETTER IDEA STORIES

Netflix

There seems to be no question that Netflix had the better idea with respect to video processing and handling. They presented what they considered to be a better idea with respect to video handling and streaming to Blockbuster, and proposed a partnership with Blockbuster. The latter was not impressed and fairly quickly rejected the offer. Netflix continued on their own and Blockbuster was on their way to bankruptcy in 2010, after they lost some $200 million in the prior accounting year. As of 2008 Netflix had a stock market value of close to $165 billion and also had some 130 million subscribers. The bottom line appeared to be that Netflix brought a better business model to the table, but Blockbuster was not buying it, preferring to stick with an old model and way of doing business. By the time they "saw the light", then CEO John Antioco was fired (2005) and Reed Hastings walked away with the victor's spoils. They were the Kings of video (DVD) subscriptions and streaming as well as expanding business areas such as its own movies and programming in a highly competitive market. Netflix had the better idea, pressed forward with it, and took risks that Blockbuster was unable to fathom.

IBM: Watson, Sr. and Jr.

IBM is a complex multigenerational story that starts roughly with Tom Watson Sr. as he took over as President in 1924. His basic idea was the next generation computer, moving ahead of so-called competitors at that time of National Cash Register and UNIVAC. This was not an easy task but Watson had the formula as well as the fortitude for success:

1. Better computers, with appropriate performance features
2. A company that was devoted to the success of their customers
3. Excellent support and marketing services

Xerox

This is one of the great success stories, based upon more than superior thinking from the creator of xerography (electrophotography) and the president and chairman of Xerox. The creator was Chet Carlson, a poor émigré from Sweden with a breakthrough idea. The president was Joe Wilson, the entrepreneur who bet his company (Haloid) on the success of xerography. The chairman was Sol Linowitz, the legal brain behind the company and its strategic partners who turned out to be Rank in the UK and Fuji in Japan.

Key companies in the U.S., namely, IBM, GE and RCA, chased Xerox trying to make an acquisition. Various deals were put forth, none accepted despite their attractiveness. The better idea carried the day, a terrific success story. Xerox above and beyond the early years has several ups and downs, but the essence of the great idea had already been seen.

And part of the story is that Xerox spawned the Palo Alto Research Center (PARC), a highly regarded research organization. New ideas for new breakthrough systems were their stock-in-trade. If you do a search on Xerox PARC, you will find that they are still alive today, and doing what they do.

Apple

Apple had many incarnations, several of which saw Steve Jobs at the helm. Ups and downs were plentiful, but in the final analysis leading to today, the company had several very successful products such as the MAC, iPod, iPad, … and Watch. The company has been one of most valuable in the past 10 years or so, and all speak with admiration regarding the "insanely" key designer and marketer, Steve Jobs.

Amazon

In some ways, this is a surprise success story. We start with the notion of selling books, and expand to selling just about everything online with record-breaking speed and responsiveness. Can it be done, considering the competition? Jeff Bezos demonstrated that it can be done by doing it. Keep expanding, even without a bottom line profit, and pushing numerous companies out of business. Eventually, you have become the richest man in the country. That is a mixture of the better idea and the relentless execution.

Facebook

When Mark Zuckerberg and friends at Harvard developed Facebook they likely had no idea that it would become a blockbuster multibillion dollar social media enterprise. Of course they built upon the basic idea, but how many people would have an interest was not foreseen. Many companies and people of note demonstrated their interest, as for example:

- A buy-in by Peter Theil
- A buy-in by Yahoo
- A buy-in by Microsoft

And how many companies today are trying to tap into the "social media" market?

Google (Alphabet)

We were all tracking the progress of Larry Page and Sergey Brin even before Google went public. So we knew the power of the search engine notion. When google went public, all of that power was confirmed, again in monetary terms. Today, Google has become a verb, as in we will "google" this or that. A better idea on top of a very good idea is a long shot, but it does happen. This is an example.

Oracle

A huge number of Department of Defense (DoD) information systems contain a database management system (DBMS). Today, the DBMS of choice appears to be Oracle, who took the lead position away from IBM's DB2. Insiders likely know how that happened, but it did, and Oracle is the up-front system of choice today. It is a better idea on top of a very good idea, as in the search engine domain.

SAS

The SAS Institute is a "home-grown" statistics system that has been carefully managed by Jim Goodnight into a $3 billion company. SAS is known for its large investment in R&D (at one point twice the industry average), an indicator of the company's belief in the better idea. The SAS Institute in Cary, North Carolina, is practically its own village, generating idea after idea in a college-like environment.

FedEx

FedEx (previously the Federal Express Corporation) is one of the largest contractors with the U.S. Government. It had the idea of overnight shipping service through one primary city, namely, Memphis, TN. This unique solution to the so-called traveling salesman and delivery service problem gave the leading position to FedEx which they have maintained for many years.

A GROUP OF GENERIC "PROBLEMS" AND "SOLUTIONS"

Blockbuster and Netflix

For problem number one, we note again that Blockbuster experienced much anxiety when visited by Reed Hastings to try to make a deal. As pointed out above, Blockbuster could not solve this quite serious problem and eventually had to file for bankruptcy. It turned out that, as of now, Netflix had the solution.

Industry and the DoD

For problem number two, we cite the many (probably in the thousands) companies that have been building systems for the military over the years. This is a difficult undertaking, but these companies have been at it for years and know how to navigate the choppy waters. The problem? The systems do not meet the performance specifications. An example? We are building an information system with a required response time, but the system will not meet that response time. This author has told two stories of this type: one experienced by Barry Boehm and the other by the author. Bottom line? Industry has taken a "systems" approach, believing that:

   a. The customer will want to avoid controversy
   b. There will always be room for a reasonable negotiation
   c. They are ultimately operating in a benign system

So in the above two cases we went to the customer (part of the system) and asked for relief from the response time requirement, and in both cases we were able to obtain such relief.

What Business Are We In?

So it turned out that the Association of American Railroads was having its yearly planning session. They asked the question that they had before, at all strategic meetings:

- What business are we in?
  And the answer came back:
- We are in the railroading business, of course
  All of that sounds reasonable, but they did not say:
- We are in the transportation business.

  We conjecture that this might well be the reason that none of the railroaders ever got into the air transport business in a serious way. What a difference just a few words might make in setting the course for the future.

  This author is told that a similar dialog was played out when Blockbuster was asked what business they were in. The answer came back: we are in the DVD rental business. They did not say, apparently, that they were in the entertainment business. So this has to do with broadening one's perspective as to the business that you are in, which leads to new ideas and new business opportunities. That leads to an activity known as innovation.

IBM and Microsoft

IBM, somewhere around 1979, was in need of an operating system for its new personal computer. They eventually engaged Microsoft to do the job for them (a long story in itself). As part of that engagement, Microsoft was legally able to offer MS DOS to every customer of the IBM PC. This became the basis for Microsoft's Windows and essentially everything going beyond that time. Why did this happen? The conjecture is that IBM did not properly value software at that time, coming from a "Big Blue" mentality. So IBM "gave away the store", paying for that bad contract agreement for years to come. And, of course, Microsoft gained substantially.

Dr. Deming and Japan and the US

Looking at potentially different futures if certain key decisions were different is part and parcel of the "what if?" world. It is another way to broaden thinking and another way to get used to considering alternatives. We might try to do this with respect to Dr. Edwards Deming, the eminent quality assurance statistician. He offered new approaches to the US automotive industry. They registered little interest until Dr. Deming was in great demand in Japan. In fact, he gets credit for bringing Japan out of losing positions in that industry and leading them into worldwide dominating positions.

Xerox PARC and Steve Jobs

Part of the Xerox PARC story is the incident whereby Steve Jobs was invited to Xerox PARC as a friendly gesture. He was shown several leading edge products and systems, which he basically "borrowed" from his host. These products and systems have been cited as the mouse, the graphical user interface, networking, windows and icons. If true, this remains a quite productive day for Mr. Jobs. Is the story actually true, or is just a part of it true and the rest of it false?

Digging more deeply into the story (by this author) has revealed that much of it is incorrect. So the ideas remain the ideas, and what happened to and with them are just part of history. But it has the apparent effect, even today, that folks are quite careful about revealing to possible competitors what it is that they are doing. You can be fighting a law suit for many years if you decide not to heed that piece of advice.

Saul Steinberg and Leasco

Back in 1960 my wife and I went off on a 10-day vacation from work and when we returned we found that my employer, Operations Research Inc., had been sold. I had a small cache of stock so that this event was of special interest to me. Of greater interest was the nature of the whole deal, and what lessons I might learn from all of it.

The company that bought us was called Leasco and was run by a gentleman by the name of Saul Steinberg. Mr. Steinberg had built the company quickly, based upon an idea, and only one idea. As part of his thesis at the Wharton School, he found out that he could buy IBM computers and lease them to customers at rates more advantageous than could IBM. So he turned that one idea into action, borrowed some money, and began buying and leasing. He tried to stabilize the company by making several acquisitions, and we were just one of them. I was intrigued by the whole adventure and entered the new world with a sense of excitement. I even met an old high school friend at Leasco who had found his way to being an executive in that company.

So here I experienced, first-hand, what a good idea can do for you and how quickly it can be monetized. It opened that door for me, and I remember the entire experience with some fondness and some degree of wanting to return to those days. Saul Steinberg is long since gone, and Leasco, I believe, was indeed stabilized by means of a critical purchase of an insurance company by the name of Reliance which became the Reliance Group.

Coke/Classic Coke, Old Taste

So there was Mr. Steinberg with a new and better idea that could be monetized quickly, and there was the worse idea, otherwise known as the Coke change of formula. You will remember that back in 1985, Coke had the brainchild of changing the basic Coke formula and taste. Good idea? Probably the executives thought it was wonderful. But the public did not buy into it and registered a considerable amount of displeasure. Nothing like your customers being very clear about what they like and what they do not like. Yes, Coke was listening and before long, the original Coke was back on the grocery store shelves.

Good Idea, Failed Follow-Through

Back in 1979, Ross Perot, who had built Electronic Data Systems, had the idea of buying Microsoft. At that time, Microsoft was a $2 million software company located in Bellevue. So he contacted Bill Gates and obtained an asking price of between $40 and $60 million. Perot thought that was too high and so he declined to make the purchase. Apparently, the offer that Perot made of between $6 and $16 million was not of interest to Gates. One might call this adventure "good idea, not consummated".

Acronym for the Individual with Potentially Better Idea

Here is an acronym for the person in a company that one wants to support because he or she has the strong potential for coming up with the right idea, at the right time.

P = PERSEVERANCE (A LA EDISON)
E = ENVIRONMENT SUITABLE FOR IDEA CREATION
T = TIME EVERY DAY TO WORK ON IDEA
I = INCENTIVES, ONE WAY OR ANOTHER
T = TEST TIME AND FACILTIES AVAILABLE (run experiments)
E = EARLY EXPOSURE TO IDEATION, ABILITY TO FAIL
WITHOUT PUNISHMENT

EXERCISES

1. Which of the dozen innovations in Table 3.1 do you find most compelling? Explain.
2. Which of the "better idea stories" do you find most compelling? Explain.
3. If you were running a company, how would you assure that it was innovative? Explain.

REFERENCES

1. Isaacson, W., *The Innovators*, Simon & Schuster, New York, 2014.
2. Eisner, H., *Thinking – A Guide to Systems Engineering Problem-Solving*, CRC Press, Boca Raton, FL, 2019.

Approaches to Problem-Solving 4

This is the point at which we identify and explore a dozen specific approaches to problem-solving. The term "specific" is relative. One specific is de Bono's lateral thinking. Another specific is Statistics-based. Yet a third specific is the Systems approach. But we will go with the flow and accept the dozen as a way to help us move in the direction of solving the problems that we might face. First, we look at the entire list, as in Table 4.1.

THE "N STEP" DISCIPLINES

In Chapter 1 we have identified a generic 5-step approach to problem-solving. This is a sparse and fundamental step approach. Beyond that, there are other "step" approaches that have been identified and used by various parties and groups. As an example, there is the well-known "8 discipline" approach,

TABLE 4.1 Twelve Approaches to Problem-Solving

1. The "N step" Disciplines
2. Reductionist/Technical Decomposition
3. Modeling and Simulation
4. Lateral Thinking
5. TSI
6. Generalized/Systems Approach
7. Design (IDEO) Approach
8. Expert Systems
9. Definitive Mathematics and Statistics-based Methods
10. DoD-suggested
11. Decision Support Systems (Software)
12. Cost-Effectiveness Analysis

originally developed and used by the Ford Motor Company. There are also the various Department of Defense (DoD)-developed and used approaches that we will see later in this chapter. All of these "step" approaches, then, become a category of approaches for our purposes in this treatise. If you like and want more detail, you might insist upon more well-defined steps. If you are comfortable with a bare-bones approach, go with the generic list as in Chapter 1.

TECHNICAL DECOMPOSITION/ REDUCTIONIST

This author's book on thinking identified the reductionist approach as a way of thinking. That certainly is and was the case. It is both a way of thinking and a way of problem-solving. It goes back to Descartes and other great thinkers such as Aristotle and Socrates. It is one of our most powerful approaches today.

In this approach, we are able to take a difficult and multifaceted problem and break it into subproblems, solve each subproblem, and then put these pieces back together to form a solution to the original problem. This cannot be done with every problem, but it can be done with many. In Chapter 1 we mentioned CIAP, the Climatic Impact Assessment Program. In fact, our approach to that real-world problem was reductionist. It was formed as basically four connected problems, connected by conditional issues where the outputs of one submodel fit into the inputs of the next submodel. Here is an explanation:

Submodel One: Compute the effluents from a fleet of postulated super-sonic transport aircraft

Submodel Two: *Given* those effluents, calculate the effects on the upper-air chemistry

Submodel Three: *Given* that new upper-air chemistry, compute the effects on the lower-air atmosphere

Submodel Four: *Given* the new lower air chemistry, calculate the effects on the human being, especially on the possible increase in the incidence of cancers

Many reductionist problems are posed as conditional probability models. That is, they are end-to-end probabilities each of which is posed as a "given" statement, such as that shown above with CIAP. Generally, we are most pleased when a reductionist approach is natural and feasible. Part of the beauty of that approach is that we can separate the problem solvers; some can be in our country, some can be overseas, some can be in industry and some can be at Universities.

MODELING AND SIMULATION

Although these two can be vastly different, both are attempts at replicating an area, or system, of behavior. As examples:

- If we want to study the weather, we build a model of it
- If we want to understand warfare, we build a model of it
- If we want to understand what happens in jungles, we build a model of them
- If we want to understand rivers and oceans, we build models of them

Simulation can be thought of as a particular form of model-building. Two basic notions for simulation are time-driven and event-driven. One simulates the behavior of domain of interest and keeps statistics on its behavior over some overall time period. Then decisions are made based upon what the statistics are telling us.

We list here below some of the software packages available to the problem solver [1]. The reader has the benefit of this rich array of capabilities:

- aGPSS
- Analytica
- AnyLogic
- ExtendSim Pro
- GoldSim
- Oracle Crystal Ball
- ProModel Simulation
- SAS Simulation Studio
- Simio Design Edition
- Simprocess
- Witness

For those that are "old school", you might remember some of the early simulation approaches for such systems as:

a. SLAM
b. GASP

    c. SIMSCRIPT
    d. DYNAMO
    e. GPSS
    f. SIMAN
    g. SIMLAB

Three areas that are especially noteworthy in terms of simulation approaches are warfare, general service and flight simulation. For the former, one sets up opposing forces and the rules that both sides must follow, and then sets the simulation in motion. The simulation will fight the war until you say "stop" or until the war is over. Then you analyze what happened by looking at stats and timelines. Better way than real world to fight the next war! And on the service side, one might imagine that the problem is to provide service from a series of ground stations to a series of satellites. The satellites are near-earth orbital and also communications-oriented. The ground stations are situated around the world at certain lat-long positions. You set the simulation in motion and begin to service the first satellite from the first ground stations. You identify the equipment needed to support that satellite and then continue on as the next important event occurs. You keep stats and timeline profiles on what is happening at each ground station in an attempt to optimize your ground station usage and design. Planning future ground stations and upgrades are a topic of importance in this latter scenario. And in the third area, flight simulation, one can fly from city to city with a simulated flight that feels very real, and can be used, according to my friends, as a surrogate instead of the real thing.

LATERAL THINKING AND (LATERAL) PROBLEM-SOLVING

This approach to problem-solving is based upon consistent Lateral Thinking, as devised by de Bono [2] and situating on this author's "thinking" list [3]. You have been digging a hole in a particular thinking and problem-solving domain for quite a long time. The metaphor is with your digging approach, you have made the hole deeper and wider. One day, you decide to start digging in another location, a lateral location. That is, you have accepted a lateral approach, a distance from where you have concentrated up to now.

A second example is that you have been researching the field of cancer and concentrating on radiation treatment for prostate cancer. You have now become an expert in that field. But then you decide to change to a different, lateral approach. That approach is chemotherapy. That is lateral choice from where you have been. And beyond that, when and if the time is ripe, you move laterally to immunotherapy and then to surgical therapy. The needle is in the haystack somewhere, but where is the haystack?

TOTAL SYSTEMS INTERVENTION

This approach is described as a "meta" method that enables creative problem-solving. It may also be viewed as a "system of systems methodology". It was developed by two well-known researchers and writers from the UK [4], Robert Flood and Michael Jackson. The former has stated that this Total Systems Intervention (TSI) is "an approach to problem solving in any organization that stands firm with the original holistic intent of systems thinking". He goes on to say that this meta-method enables the problem-solver to select the kind of problem first and then the specific method that best applies to that problem. This acknowledges that some methods are better or worse in terms of their applicability to the problem at hand. So it makes sense to be able to take that extra step of selecting the domain of the original problem.

This was a clever idea and is best illustrated by an example or two. An organization has many different possibilities for problems. We will call them problem domains. Here is an example of such domains for an arbitrary organization or enterprise:

a. Financial strength and accounting
b. Strategic planning
c. Project execution
d. Ability to market effectively
e. Human resources
f. Internal functions and departmental operations

Let us assume that it is the strategic planning area that needs help. Given that selection, management chooses a decision support system (DSS) as the method that would be most useful in solving a problem in that domain. Another example might be that the enterprise has a problem in managing its projects. So they look for a method that will be able to help with project management. A package of software (like Microsoft Project) might be helpful in this regard.

GENERALIZED/SYSTEMS APPROACH

This method emphasizes generalizing and overall use of the systems approach. The reader is reminded of the elements of the systems approach as listed below [5]:

1. Establish and follow a systematic and repeatable process
2. Assure interoperability and harmonious system operation
3. Consider alternatives at various steps of design
4. Use iterations to refine and converge
5. Create a robust and slow-die system
6. Satisfy all agreed-upon user/customer requirements
7. Provide a cost-effective solution
8. Assure system sustainability
9. Use advanced technology, at appropriate levels of risk
10. Consider all stakeholders and their concerns about the system
11. Design and architect for system integration
12. Employ systems thinking

An integral part of this method is addressing a problem area by generalizing. The author has found that this idea is useful in many cases and situations. As an example, we look at the matter of defining one's business in the context of strategic planning. The question is:

"What business are we in?"

Your answer is quite revealing and tells us whether or not you are interested in systems thinking.

So let us look at a final example of this type of thinking. We will assume that you are a builder of airport surveillance radars for the Federal Aviation Administration (FAA). You have an adequate share of that market, but you have now decided to somehow increase your revenues. You ask the strategic plan question:

"What business am I in?

One answer, an accurate one, is – "I'm in the airport surveillance radar business."

That answer is not conducive to systems thinking. For our purposes, we provide another answer to the question, which is:

"I'm in the radar business"

That broader answer prompts us to look at other types of radars, such as:

- Air route surveillance radars
- Harbor radars

- Airborne radars
- Police radars
- Any other types

This look will provide a forum for your strategic plan and probably lead you to new markets within the field of radars. Systems thinking has served its purpose and has helped you to solve a problem.

The classical systems approach looks at problems in the context of the system or systems of which they are a part. For example, you are building a system, and you ask yourself – is the customer part of the system? The answer is definitely "yes".

DESIGN APPROACH

This refers to the specific solution as part of Tim Brown's book on "Change by Design" [6]. The reader will recall (see Chapter 2) that Brown was the author of this "seminal" text in this [6] arena and Tom Kelley has led the premier company IDEO. Tim Brown has been quoted as clarifying [5]:

> Design thinking is a human-centered approach to innovation that draws from the designer's toolkit to integrate the needs of people, the possibilities of technology, and the requirements for business success

Design thinking [7] has been contrasted with traditional problem-solving in that the latter is more linear and structured, usually working from a given fixed data set. Design thinking is more user-oriented, with divergent thinking and the use especially of prototypes. The reader is urged to go back to Chapter 2 for more information.

EXPERT SYSTEMS

Artificial Intelligence has many facets to it, with one being especially useful for a certain type of problem-solving. That part has been known as "expert systems". These systems provide expertise in a variety of areas, those which have been populated with facts and figures in a particular field [8]. An example is the field of medical diagnoses and the problem that is potentially being solved is that of figuring out what might be medically wrong with a given patient. One enters the system (a piece of software) with a list of symptoms and the output soon appears – what might be wrong, what the likelihood is and relevant commentary.

An example of the strength of an expert system is the Watson system. This was a system produced by IBM and put to the task of beating the Jeopardy champions in a head-to-head match. The machine won, and gained notoriety from the win and the overall notion of machine over man. Watson also beat the prevailing chess champion (Spassky) back in 1977.

How does an expert system work? It basically has only three components: (a) a user interface, (b) a knowledge base and (c) an inference engine. The inference engineer queries the knowledge base and sends answers back through the user interface. The inference engine does all the "heavy lifting", so to speak, while the knowledge base contains huge amounts of information.

Here below is a short list of Expert Systems that are available as software:

XCON
MUDMAN
Authorizers Assistant
ExperTAX
Consultant
Financial Advisor
Plan-Power
XCEL

DEFINITIVE MATHEMATICS AND STATISTICS-BASED METHODS

There are quite a few well-defined quantitative methods that can be impressed into service when the problem is right. These are well-known and also easy to access. They are mathematics-based, with a short list as below:

- When you are looking for a quantitative minimum or maximum – use calculus, and a special form known as calculus of variations
- When you are looking for a two-way table – use a spreadsheet
- When you are looking to optimize some functions – use optimization software
- When you are looking to write a small program – use Excel
- When you are looking to graph or plot some data – use Excel
- When you are knee-deep in data – use analytics to make sense of it all

- When you have a complex linear programming problem – use appropriate software
- When you are trying to find the local maxima or minima of a complicated function, or optimize that function – possibly the method of Lagrange Multipliers is the way to go
- When you have a system flow problem – use Forrester's System Dynamics Method

The long and short of it is that there are many problems that need to be addressed by definitive mathematics. The trick, of course, is to know what mathematics to use when. Also, various mathematical approaches have been embedded in software so if you use that software, you are using an approach but may not know it. An example is the use of optimization software that uses Lagrange multipliers but that fact is not known by the user. Nonetheless, use of the software is totally legitimate.

Statistics-Based Approach

Statistics uses probability theory as its base, so when your problem can be stated as a probability, you can reliably go to statistics as an approach. Statistics, as a discipline, has a rich assortment of submethods, if you will. An example of such submethods is shown here in the list below:

- Probability theory
- Correlation analysis
- Estimation theory
- Sampling theory
- Hypothesis testing
- Least squares fitting
- Chi-square testing
- Specific distributions (e.g., Poisson, Binomial, etc.)

DoD-SUGGESTED

The Military, one could expect, has addressed this issue in all their services. Looking at two of them, the Army and Air Force, respectively, we see the two lists below, [9,10]:

1. Recognize and define the problem
2. Gather facts and make assumptions

3. Define end states and establish criteria
4. Develop possible solutions
5. Analyze and compare possible solutions
6. Select and implement solution
7. Analyze solution for effectiveness

1. Clarify and validate the problem
2. Break down the problem and identify performance gaps
3. Set improvement target
4. Determine root cause
5. Develop countermeasures
6. See countermeasures through
7. Confirm results and processes
8. Standardize successful processes

Some commentary on the above two approaches follows:

- One approach is seven steps and the other is eight
- Despite that difference, the two approaches are quite similar
- The Air Force requires that the problem be "validated", which makes sense; we do not want to be spending a lot of time, energy, and dollars on a nonproblem
- Countermeasures as used by Air Force is in essence the same as "solution"
- The Army criterion of "effectiveness" requires that the problem-solver accept the notion that cost-effectiveness will be part of the overall approach, i.e., we are committed to finding the most cost-effective solution [11]
- The Air Force declares the importance of having a standardized successful process

The two above approaches can be compared with what I will call the General Staff approach, as listed below:

1. ID (identify) the Problem
2. Facts and Assumptions
3. Generate Alternatives
4. Analyze the Alternatives
5. Compare the Alternatives
6. Make and Execute Decision
7. Assess the Results

As a footnote to the multistep approach to defining a problem-solving process, we take here a turn toward academia and to our foremost guru, Peter Drucker. Although the man himself has passed on, his deep and brilliant work lives on, an example of which was what he had to say about a five-step problem-solving process. Taking the "road less traveled", Drucker defines the five steps as [12]

1. Develop a rule or principle regarding this type of problem
2. Write a specification as what the solution must satisfy
3. Determine what will fully satisfy the above specification
4. Define specific actions to carry out the solution
5. Monitor feedback that tests the validity and effectiveness of the solution

How is That for Out-of-the-Box Thinking?

Predictably, Drucker comes at problem-solving in his own way and often quite differently from the rest of us mere mortals. For example, Drucker's first step worries about the generality of the problem and implicitly wants to "solve" (or at least state) the general problem. Then he is concerned about specification for the solution. Finally, he is worried about the validity of the solution. Let us think long and hard about Drucker's approach that is so different from the mainstream.

DECISION SUPPORT SYSTEMS

These systems, implemented in software, represent a powerful way of solving problems. A usual scenario is that a group gets together, around a conference table, to address and solve a problem. The group has a leader who gets the DSS up and running. The leader declares that the group purpose is to reach consensus on the problem and its solution. The expectation is that reaching consensus on the problem will be easy; doing the same on the solution will be difficult since the claim is that there is not enough time to truly analyze a set of alternatives. A reasonable expectation in this area is that consensus is reached on the alternative solutions that will be considered. Then the meeting is over and the participants go off and spend quality time (possibly weeks), with their constituents, analyzing the alternatives. After an agreed-upon time, the group comes together again to examine the results to date and engaging in the new consensus-gaining process. The new focus is the set of suggested and analyzed alternatives. What is the consensus, Alternative A, B or C?

Listed below we have a citation of available DSSs:[13]

Decisions
Airfocus
QuickScore
Style Intelligence
Yonyx
Ibi
Qlik Sense
SAP BusinessObjects
Wolfram Mathematica
TIBCO Software
Salesforce Analytics Cloud
EIDOS
Loomio
FactoryTalk Innovation Suite

In today's world, problem-solving can be thought of, in many cases, as finding the right software that will use the right modeling capability, or simulation, or mathematics, or statistics. How do we know what the right choices are? That is what age and experience are all about. That is when it helps to be able to say – "I've seen something like this before".

COST-EFFECTIVENESS ANALYSIS

The final item on our list of problem-solving approaches is cost-effectiveness analysis. This can be traced back to the military in the 1960s when they produced the so-called WSEIAC report [14]. This report defined three measures that were important with respect to measuring the effectiveness of military systems, namely,

1. Availability
2. Dependability
3. Capability

Their precise definitions were as follows:

"**Availability**. A measure of the condition of the system, at the start of a mission, when the mission is called for at an unknown (random) point in time.

"**Dependability**. A measure of the system condition during the performance of the mission; given its condition (availability) at the start of the mission."

"**Capability**. A measure of the results of the mission; given the condition of the system during the mission (dependability) and **cost-effectiveness** is the value received (effectiveness) for the resources expended (cost)"

Soon thereafter, the DoD basically declared that they were looking for, and insisting upon, "cost-effective" systems, and one needed to demonstrate that they were in tune with that orientation. That being the case, it was then necessary to precisely define both costs and effectiveness, the latter being the most difficult. That perspective has carried forth to today and thus becomes the final item on our list of problem-solving approaches.

EXERCISES

1. What are your "top three" choices of approaches to problem-solving? Explain.
2. In three pages, write an example of how you would use TSI.
3. In three pages, use the IDEO approach to design a new product. Be specific about the product.

REFERENCES

1. Swain, J., "Simulation takes over; Reality is for sissies", *ORMS Today*, 44: 38–49. October 2017, INFORMS.
2. Bono, D., *Lateral Thinking*, Harper & Row, Manhattan, NY, 1970.
3. Eisner, H., *Thinking – A Guide to Systems Engineering Problem Solving*, CRC Press, Boca Raton, FL, 2019.
4. Flood, R., and M. Jackson, *Creative Problem Solving – Total System Intervention*, John Wiley & Sons, Chichester, 1991.
5. Eisner, H., *Topics in Systems*, Mercury Learning and Information, 2013.
6. Brown, T., *Change by Design*, Harper Collins, New York, 2019.
7. See https://designthinking.ideo.com. Design thinking is a human-centered approach to innovation that draws from the designer's toolkit to integrate the needs of people, the possibilities of technology, and the requirements for business success.
8. See https://en.wikipedia.org/wiki/Expert_system.

9. See www.armystudyguide.com.
10. See www.af.mil/News.
11. See www.armyupress.army.mil/Journals/NCO-Journal/Archives/.
12. Sublimeyourtime.com/peter_drucker.
13. See en.wikipedia.org/wiki/DecisionSupportSystem.
14. WSEIAC, Weapon System Effectiveness Industry Advisory Committee, Chairman's Final Report, AFSC-TR-65-6, January 1965.

Think Tanks and Problem-Solving

5

This chapter explores a few selected labs, think tanks and federally funded research and development centers (FFRDCs) whose main job is to think about ways of solving problems on behalf of their customers. By and large, they do an exceedingly good job and they are at the leading edge of research and development for the country. We have much to learn from them, if we are able to observe and understand what they are up to. There are restrictions, however, as to what they are able to tell us. Such is life in the fast lane.

In this author's book on "Thinking", there is a short discussion of three of the above, in particular Bell Labs, Xerox PARC and the Applied Physics Lab at Johns Hopkins University. We move on, in this book, to other citations of note within the world of Labs, Think Tanks and FFRDCs.

FEDERALLY FUNDED R&D CENTERS

MITRE has produced a primer to explain the nature and role of FFRDCs [1]. These organizations have supported the government for some 70 years "by developing transformational capabilities in defense, transportation, energy, civil agency administration, homeland security, atmospheric sciences, science policy and others". Assistance is provided by way of research and development, study and analysis, and/or systems engineering and integration. These centers operate as not-for-profit organizations.

SYSTEMS ENGINEERING RESEARCH CENTERS

This federally funded activity is devoted to the technical area of systems engineering. It is located at the Stephens Institute of Technology in Hoboken, New Jersey. Government sponsorship is now provided at the Assistant Secretary level in the Office of the Secretary of Defense. The general intent is to have these Centers self-sufficient after a suitable period of time.

Systems Engineering Research Centers (SERC) is actually called a University-affiliated Research Center whose mission is three-fold:

a. To enable collaboration among several Systems Engineering (SE) research organizations
b. To accelerate SE competency development
c. To transform SE practice through creation of innovative methods, processes and tools

Members of SERC are ever present in the world of SE and seem to be interested in all meetings and contacts so as to satisfy their mission. They readily share their time and information, which is quite important in this domain. This SERC also claims their vision as follows:

- "To be a national resource that is able to meet systems challenges of national and global significance by means of systems research" [2]

Included among the areas of particular research interest are the following:

- Enterprise and system of systems
- Mission engineering
- System of systems modeling and analysis
- Trusted systems
- Systems security
- Systems engineering and systems management transformation
- Tradespace and affordability
- Systems engineering for velocity and agility
- Human capital development
- Research incubator

A prior version of the SERC was won in competition by the University of Maryland, with the leader in that endeavor being Dr. John Baras. That SERC has been considered to be a major success on its own as well as in terms of setting standards, goals and objectives.

THE MITRE CORPORATION

MITRE was initially established (1958) to support the Air Force's ESD (Electronic Systems Division) in Bedford, Massachusetts [3]. It continues to work with the ESD although its activities have expanded substantially over the years, to the advantage of a huge number of satisfied clients. From my experience, MITRE appeared to be very careful in terms of its business practices which included farming out work to competent subcontractors. One of my interactions in this regard was with a company that had a subcontract dealing with FAA software issues. MITRE appeared to be pleased to bring this company aboard, thus sharing revenues and responsibilities.

At the time of this writing, MITRE had revenues of $1.734 billion and 8,425 employees. MITRE has also placed itself in the position of managing FFRDCs that are supporting various government agencies. In this regard, MITRE has direct connections with the following centers, along with a listing of center sponsorships:

MITRE Centers	Sponsor
National Security Engineering Center	Department of Defense
Center for Advanced Aviation System	Federal Aviation Administration Development
Center for Enterprise Modernization	Internal Revenue Service and Department of Veterans Affairs
Homeland Security Systems Engineering	Department of Homeland Security and Development Institute
Judiciary Engineering and Modernization	Administrative Office of the U.S. Courts Center
CMS Alliance to Modernize Healthcare	Centers for Medicare and Medicaid Services
National Cybersecurity FFRDC	National Institute of Standards and Technology

MITRE has been well known for its excellence in work results and its careful management. These have been landmarks of the organization, and indicators of why and how they have done so well over the years. In this regard, some of the awards they have received have included the following:

MITRE Awards over the Years

- A North American Most Admired Knowledge Enterprises (MAKE) Award
- The Secretary of Defense Medal for Outstanding Public Service
- The Air Force Association's Von Karman Award for contributions to the field of engineering and science
- A trophy for activities in conceptualizing, developing and implementing a cornerstone capability for the future of the national airspace system

In 2015, Forbes Magazine declared MITRE to be one of America's best employers.

THE AEROSPACE CORPORATION

The Aerospace Corporation [4] was established in 1960 to provide technical capabilities in support of all space mission for Military, civil and commercial customers. It serves as the FFRDC for national security space and as such works closely with the Missile Systems Center and the National Reconnaissance Office (NRO) to provide objective technical analyses and assessments so as to achieve high levels of performance and reliability. It has some 3,600 employees, with headquarters in El Segundo, California.

The company is basically composed of Groups and Centers, as listed below.

Space Systems Group. Working with the Air Force and industry it develops military satellites and advanced national security satellite systems. Domains include communication, surveillance, weather, and navigation.

Engineering and Technology Group. Major science and engineering group consisting of some five divisions, cross-cutting program with technical analysis.

National Systems Group. Provides systems engineering and integration support for NRO space programs.

Defense Systems Group. Provides technical support to all national security space programs, in areas such as
a. Mission planning
b. Requirements analysis

c. Architecture formulation
d. Strategic planning
e. Systems acquisition support

Civil Systems Group. Supports developing and operating civil space systems for agencies such as NASA and NOAA, as well as commercial systems and their operations.

Center for Orbital and Re-entry Debris Studies. As the name implies, focuses on what happens to space debris.

Center for Space Policy and Strategy. New in the year 2000, serves as a center of excellence for overall space policy

Recent news reported by the Aerospace corporation in 2020 included:

- CubeSats are getting closer; proximity operation
- Infrared temperature scanners keep employees safe
- The solar gravity lens; a three-decade journey
- SpaceX's Falcon 9 sends space force's GPS III-3 to orbit
- Exploring the future of space travel
- Continuous production agility in action
- AEHF-6: space force's first launch
- Building the modular future of space

All of which gives one a snapshot of activity related to this highly productive company.

THE RAND CORPORATION

RAND [5] is a research organization that "develops solutions to public policy challenges to help make communities throughout the world safer and more secure, healthier and more prosperous". It is a nonprofit committed to the public interest, with some 1,850 employees.

RAND was created in 1948 by the Douglas Aircraft Company. This was a response to perceived problems of significance in the US at that time. We needed "think tank" types of people to tackle and solve these problems, which included:

- Healthcare
- National security
- Nuclear arms agreement
- Social welfare

Since then, one might observe an expansion into such diverse areas as:

- The digital revolution
- Homeland security
- Civil and criminal justice
- Labor markets
- Education
- International policy
- Crisis management
- Long-range planning
- Science and technology
- Space policy

RAND has been given credit for coming up with the doctrine of mutual assured destruction and also the beginnings of the entire field of systems analysis, as practiced initially by Robert McNamara and his Whiz Kids during the Kennedy administration. It was also given credit for its role in the SAGE system, especially with respect to state-of-the-art software development for the new computers in 1955. The SAGE system was designed to solve one of the most important national problems (early threat detection and air defense) in that time frame.

RAND is currently broken into nine divisions, addressing key problems and issues of selected countries. These include as follows:

- Homeland Security Operational Analysis Center
- Army Research Division
- Education and Labor
- Healthcare
- National Security Research
- Social and Economic Well-Being
- Project Air Force
- Europe
- Australia

ANSER

Relatively small, Anser [6] still serves as a Federal Contract Research Center, mainly for the Air Force. It has some 500 employees and is located in Falls Church, Virginia.

Areas of focus for ANSER include:

a. Homeland security
b. Operation of the homeland security institute
c. Advance technology
d. Threat analysis and mitigation
e. Forensics
f. Antiterrorism
g. Chemical/biological defense

IDA

IDA [7], the Institute for Defense Analysis, focuses on challenging national defense issues that require intense analysis and suggestions for solutions and actions. It is relatively small and is headquartered in Alexandria, Virginia.

Areas in which IDA supports its customers include:

Systems and Analyses Center Research. This research tends to deal with:
a. Weapon systems
b. Tactical doctrine
c. Intelligence
d. Advanced analytics
e. System costs

Science and Technology Policy Institute. As the name implies, this group deals with policy matters and interacts with the Office of Technology Policy and the National Science Foundation as well as several civil agencies.

Center for Communications and Computing. Areas of interest and analysis for this Center include:
a. Encryption and cryptology
b. High-performance computing
c. Advanced algorithm and mathematics
d. Computer networking
e. Communications security
f. Use of large data sets
g. Cybersecurity

CNA

This group, the Center for Naval Analysis [8], has been dealing with operations research matters and problems of the Navy for about 75 years. It was a pioneer in this domain, working on such issues as:

a. National security
b. Defense
c. Education
d. Homeland security
e. Air traffic management
f. Naval warfare

They emphasize four "seminal research projects":

1. A capstone strategy series
2. An integrated ship database
3. A population representation
4. A retirement calculator

They have highlighted the following as especially significant areas of study:

a. The modern information environment
b. A plan for the department of the Space Force
c. CNA case studies; an inside look at great analysis

They have also taken the time articulate "How" they do their research with the following ground-rules:

a. Maintain absolute objectivity
b. Apply imaginative and innovative techniques
c. Gain a thorough understanding of issues
d. Are process-driven and results-oriented
e. Are open, direct and clear

This is useful in terms of setting approaches to problem-solving by an institution in distinction to an individual.

ENERGY NATIONAL LABS

The U.S. Department of Energy National Labs and Technology Centers, executing the DOE mission, consists of some 17 FFRDCs. This system of national labs grew out of post-WW II needs in such areas as nuclear policy and materiel, radar and the atomic bomb. The labs represent one of the largest scientific research activities on the world. Some of the more familiar "parts" of the labs include:

- The Lawrence Berkeley Lab
- The Oak Ridge Lab
- The Argonne Lab
- The Brookhaven Lab
- The Fermi Lab
- The Los Alamos Lab
- The Lawrence Livermore Lab
- The Sandia Lab

This massive set of labs should be considered a national treasure in the overall field of energy. As such it tends to be well-funded, with internal issues of fund allocation and program priorities from time to time.

EXERCISES

1. If you were managing a think tank, in three pages, describe the management principles you would use.
2. Explore three areas of research likely to be embraced by the SERC.
3. Why do we need the eight national energy labs cited at the end of this chapter?

REFERENCES

1. A Primer, The MITRE Corporation, see www.mitre.org.
2. Systems Engineering Research Center, see sercuarc.org.
3. See https://en.wikipedia.org/wiki/MITRE.

4. The Aerospace Corporation, see aerospace.org.
5. The RAND Corporation, see rand.org.
6. The ANSER Corporation, see anser.org.
7. IDA, The Institute for Defense Analysis, see ida.org.
8. CNA, The Center for Naval Analysis, see cna.org.

Specific Problems and Their Solutions

6

There are numerous (possibly countless) types and kinds of problems. Therefore, there is no single solution to even a large number of problems. There are only good and bad (and in between) approaches to solving various kinds of problems. In this chapter, we select a few kinds of problems and suggest approaches to solutions and reasons for these approaches. In some cases we provide a solution and try to generalize from that solution. So this is a suggested approach chapter, one that is a possible pointer toward other solutions.

The problems addressed here are listed in Table 6.1 below.

TABLE 6.1 Selected Problem List

1. Coin Weighing
2. Types of Algorithms
3. Kalman Filter
4. Minimum Computer Steps
5. The Matchstick Problem
6. Efficient Message Coding
7. The River-Crossing Problem
8. Parameter Dependency Diagramming
9. Optimal Search
10. Laplace and Fourier Transforms
11. Lagrange Multipliers
12. Allocation of Requirement Errors
13. A Walk in the Park

THE COIN WEIGHING PROBLEM

The overall problem is posed as:

- You have twelve coins, all of equal weight but one. That one is either heavier or lighter than the others. You also have a balance scale such that you can weigh some number of coins against some other number that you choose. How are you able, in three weighings, to find the coin that is different and whether it is heavier or lighter than the others?

You might be inclined to start out by measuring six coins against six coins but I will save you some trouble. That does not work. How do you know that? You do not. But the solution lies on the path where the first weighing is four coins against four coins. As we are thinking, step-by-step, we ask about finding the first weighing. Our thought is to maximize the information we get from that first measurement. Four coins by four coins by four coins gives us more information than does the six by six. So let us do that and we find that there are three possibilities.

A. The right side is heavier B. The left side is heavier C. Both sides are of equal weight

The right side has four coins $s(1)$, $s(2)$, $s(3)$, $s(4)$. See logic and conclude that bad coin is on right side and is heavier.

We leave the rest of the solution to the reader. But if you are in trouble, consult the author's book [1] where you will find a step-by-step solution.

TYPES OF ALGORITHMS

An algorithm is a procedure, usually a series of steps, that solves a problem. If we look it up from a college course description, we find that it usually shows up as a computer science class. And there, a definition is:

"This course represents an introduction to the techniques for designing efficient computer algorithms and analyzing their running times. General topics include asymptotics, solving summations and recurrences, algorithmic design techniques, analysis of data structures, and introduction to NP-completeness".

Various folks have taken the time to categorize types of algorithms [2], with many of these types listed below. We see what a rich topic this is in relation to various types of thinking.

a. Combinatorial algorithms
b. Optimization algorithms
c. Numerical algorithms
d. Database algorithms
e. Distributed system algorithms
f. Operating systems algorithms
g. Graph algorithms
h. Network flow and theory algorithms
i. Sequence algorithms
j. Search algorithms
k. Karmarkar's algorithm

So we can see that there is no dearth of sources of algorithms that address a variety of problems. Many of these algorithms can be downloaded from convenient websites. The "thinking" that went into the development of these algorithms may be minimal, or it may be extensive. In most cases, no new thinking need to be done by the user. Just follow the instructions. The last on the above list, from Karmarkar, is in the domain of linear programming.

THE KALMAN FILTER

For our purposes, since there is a fixed and well-known procedure for a Kalman filter, we may take it as an example of an algorithm. Some years ago, this author had occasion to use the Kalman filter in his work environment. This is a filter that carries out linear quadratic estimation. One takes a series of measurement over time and follows with several calculating steps, which produce estimates of unknown variables that are better than a single estimate alone. Roughly, here are the steps to be followed:

Step One: estimate the a priori state vector, of dimension $(m \times 1)$
Step Two: estimate the covariance matrix, where $w = W(1) =$ measurement noise of dimension $(n \times m)$
Step Three: determine the measurement matrix for the system, which is an $(n \times m)$ matrix relating the measurements to the state variables
Step Four: estimate the new covariance matrix

Step Five: make new measurement of state variables
Step Six: update the covariance matrix
Step Seven: calculate the new values of the state variables

THE MINIMUM COMPUTER STEP PROBLEM

This may be construed to be an algorithm in that it defines the three steps necessary to write any computer program [1]. These three steps are the direct sequence, the decision and the loop. The reader is requested to check this out with a few google searches.

THE MATCHSTICK PROBLEM

We pose a "matchstick" problem, as follows:

We see six matchsticks lying on a table forming the Figure 6.1 in the text. The problem is to move three of these matchsticks (only three) to form four equilateral triangles. Move when ready. We now observe that an hour has gone by and the reader does not have a solution. So we will provide one here.

FIGURE 6.1   Matchstick problem.

We did not specify that the problem may be addressed by moving from two to three dimensions. Thinking long and hard will not help as long as we are staying in two dimensions. If we move to three dimensions, we take the bottom three matches and stand them up to be near vertical, with edges converging to a point. Lo and behold, now we are able to construct the required equilateral triangles. We note that we are "thinking outside the box" as suggested in the author's previous book. In particular, we are "expanding the dimensions", literally.

THE EFFICIENT MESSAGE CODING PROBLEM

Some years ago, this author was engaged in teaching a course in information theory. As part of that course, he ran into that posed problem – if we have a given message, and we wish to encode it with zeros and ones, we can follow a procedure that has been shown to be efficient.

Let us take a simple case where we have four messages (A, B, C, D), with their given probabilities as below:

MESSAGE	PROBABILITY	BINARY CODE	HUFFMAN CODE
A	½	00	0
B	¼	01	10
C	1/8	10	110
D	1/8	11	111

Two codes are shown, the Binary Code and the Huffman Code. Although the Binary code looks very efficient, it turns out that the Huffman code is, in general, more efficient. Can you show that in the particular case of the above four messages? What are you thinking when you confront this type of problem?

THE RIVER-CROSSING PROBLEM

There are several variations on the so-called river-crossing problem. We will briefly look at just one of them as reflected in the wolf, goat and cabbage scenario. Here the farmer goes to the market and buys a wolf, a goat, and a cabbage (why he does this is unknown). He comes to the edge of a river which is on his way home. How does the farmer cross the river and bring home his purchases intact?

If unattended, the wolf will eat the goat, or the goat will eat the cabbage.

What is the first action? The solution calls for taking the goat across the river. This leaves the wolf with the cabbage, which is not a worry. The farmer returns, having left the goat on the other side. He is confronted with the wolf and the cabbage. What next?

Next he can take the wolf over, but when he returns, he does so by taking the goat back with him. When he returns, he now has with him the goat and the cabbage.

Next take the cabbage over, leaving on the other side the wolf and the cabbage.

Next he returns, picks up the goat and takes it with him to the other side. Done deal, and the key step was to take the goat back with him when he returned from the other side. Was this obvious to the reader? Perhaps, but it is clearly nonlinear thinking. Most of the river-crossing problems are highly nonlinear and require "out of the box" thinking.

PARAMETER DEPENDENCY DIAGRAMMING

Parameter Dependency Diagramming grew from a need to model very large-scale systems, to include systems of systems. An early application area was the National Aviation System (NAS) under a contract that this author had with the Federal Aviation Administration (FAA) [3]. The basic idea is that even though we may not know the mathematical relationships between the key variables, we still wish to capture their dependencies. The Parameter Dependency Diagramming (PDD) does this, with some precision. We will explore this notion, first with respect to the NAS, and then with respect to a typical radar system.

The National Aviation System

The formulation of a NAS model was based upon certain perceived needs within the FAA, namely [3]:

a. Lack of a unified structure of the NAS within the FAA
b. Diversity of existing models

c. Noncomparability of outputs
d. Complexity of inputs
e. Excessive time requirements

It was decided that the best approach involved characterization by thirteen attributes (some call them Measures of Effectiveness – MOEs), as listed below:

1. ATC Availability
2. Airport Capacity
3. Airport Delay
4. Airspace Capacity
5. Airspace Delay
6. Trip Time
7. Energy Utilization
8. Service Availability
9. Noise
10. Pollution
11. Security
12. Safety
13. Costs

Each of the above, therefore, was represented by a "submodel". Due to the submodel complexity, the PDD was used to establish the framework for each submodel as well as the overall model.

In terms of thinking, this approach can be seen to be reductionist. The problem is broken down into pieces, in this case the set of attributes and their corresponding submodels. Then each submodel addressed in whatever detail is necessary. Many of these submodels are of considerable difficulty (e.g., the delay submodels). Many are more straightforward (e.g., airport capacity). This is truly a model of submodels with the structure provided by the parameter dependency diagramming.

A Radar Submodel

We will continue by illustrating the PDD for a simple radar example. The key dependencies are represented by five variables, namely:

a. The detection probability, $P(d)$
b. The false alarm probability, $P(\text{fa})$
c. The output voltage (V)

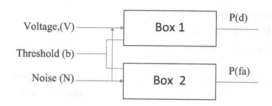

FIGURE 6.2 PDD and formulae for radar example.

d. The noise power (N)
e. The detection threshold (b)

We will be detecting the signal (output voltage) in the presence of additive Gaussian noise. The detection probability is the area under the signal plus noise distribution to the right of the threshold, and the false alarm probability is the area under the noise alone distribution to the right of the threshold. The PDD and formulae for these relationships are shown in Figure 6.2 above.

AN OPTIMAL SEARCH

Many situations call for searching for one thing or another. Radars are searching for targets in a local environment for civil aviation, or in a global environment as in the national missile defense scenario. We can look at search in a very local case, as in the problem posed and discussed below.

Let us assume that keys have been tossed into two tin cans such that the desired key is twice as likely to be in the red can as in the blue can. This is translated into [4]:

Prob(red) = 2/3 and Prob(blue) = 1/3

We further assume that with respect to detection,

$$P(t) = 1 - \exp(-t)$$

where $P(t)$ is the probability of detection if searched for time t, given that the search is carried out in the can with the desired key in it. It can be shown [4] that an optimal allocation for a total available time of 4 hours is:

- Allocate a total search time to the red can of $2 + (1/2)\ln 2$
- Allocate a total search time to the blue can of $2 - (1/2)\ln 2$

Although this is a simple example there are many search patterns and results that are very important. In this regard, the search for Bin Laden took many years and a tremendous amount of effort. Searches persist today in the same context and related to national security. Searching for incoming missiles in a national defense scenario is also of extreme importance.

LAPLACE AND FOURIER TRANSFORMS

Today's electrical engineer is well-versed in linear systems analysis involving Laplace and Fourier transforms. These are powerful techniques and serve the analyst well in circuit and system analysis. The concept is that we convert to a domain where the functions are multiplicative, a much more desirable place to be. The reader is pointed to the many texts available on these subjects.

LAGRANGE MULTIPLIERS

Another method for looking for an optimum is that of Lagrange multipliers. That method is used, for example, to show that entropy (H) is maximum when all the probabilities are equal, namely, $1/n$. The procedure is relatively straightforward. The following function is constructed [1]:

$$H + \text{lambda}\left\{ p(1) + p(2) + \cdots + p(n) \right\} - 1$$

Followed by differentiation with respect to the $p(i)$, which leads to the set of equations

- $\left[\log p(1) + 1 \right] + \text{lambda} = 0$
- $\left[\log p(2) + 1 \right] + \text{lambda} = 0$
- ...
- $\left[\log p(n) + 1 \right] + \text{lambda} = 0$

which implies that $\log p(i) = \text{lambda} - 1$ for all $p(i)$ and therefore $p(1) = p(2) = \ldots p(n)$ and $p(i) = 1/n$

ALLOCATION OF REQUIREMENTS ERRORS

Here we take a brief look at allocating errors in a system where such are prevalent and important. Such a system might be a telescope in space, where there are three significant error sources that dominate that scenario.

We will assume that the overall pointing error is sigma = 10 arc-seconds and that the three error sources are:

a. The spacecraft control system error = error (1)
b. The instrument control system error = error (2)
c. The optical path distortion error = error (3)

These errors are estimated as error (1) = 6 arc-seconds, error (2) = 6 arc-seconds and error (3) = 5.3 arc-seconds. We note that the errors must obey the mathematics attendant to error analysis, namely:

Sigma squared total = sigma (1) squared + sigma (2) squared + sigma (3) squared

which leads to the numerical values of

$$(10)^2 = (6)^2 + (6)^2 + (5.3)^2 = 36 + 36 + 28 = 100$$

This is a simple example but it is of major significance in the systems engineering world where we are both trying to define and control errors.

A WALK IN THE PARK

When you take a walk in your local park, you often see a chain hanging from two posts, enclosing a patch of grass. The chain looks like it takes the shape of a parabola. If you look up, you will see freely hanging overhead power (high voltage) lines, also looking like parabolas. However, it can be shown that they are not parabolas, they are catenary-shaped. The graph of this form is the hyperbolic cosine function. This function has been known as far back as 1670 (Robert Hooke) and are solutions to Maxwell's equations as well. The spider sets forth this same shape in its web. From the small to the large – surprising! Who knew the spider was that clever, choosing an "optimal" shape?

The Bottom Line

This chapter has pointed us to a variety of problem areas, each of which has a different approach to solution. We would be pleased to have a "universal" or "dominant" approach, but such is not the case. We have to "think" our way through each and every problem, and we find the approach that is suitable. Such is life. The bottom line is to keep in touch with the literature and be prepared to do a google search for solution approaches from that medium.

EXERCISES

1. Complete the coin weighing solution.
2. In three pages, cite areas in which you would benefit from knowing how to apply search theory. Explain.
3. Show how the walk in the park leads to the catenary observation.

REFERENCES

1. Eisner, H., *Computer-Aided Systems Engineering*, Prentice-Hall, Englewood Cliffs, NJ, 1988.
2. See https://en.wikipedie.org/wiki/ListofAlgorithms.
3. "National Aviation System Modeling Effort – Overview", March 18, 1977, ORI, Inc., 1400 Spring Street, Silver Spring, MD 20910.
4. Stone, D. F., *Theory of Optimal Search*, Academic Press, New York, 1975.

Great Thinkers and Problem-Solvers 7

This chapter explores the work of some of our great thinkers with an eye toward *how* they thought. It is not focused on what they thought about, since that is mostly what we all know, from their results and their writings. But the more interesting idea, for purposes of this treatise, is the "how" part of the equation. We are hopeful that we can better understand the "how" so as to try it for ourselves in our problem-solving today.

As an example, we know what Einstein thought about and the wonderful results that he achieved. But what is lesser known is that the "how" of his thinking preferred the process of "visualization". This was not the only way he thought but it was an especially useful one. To put this in another way, if Einstein had not been able to visualize his thoughts, he very likely would not have been able to obtain the results that he did.

So we are in search of the "how", even though that is a longer road to travel, and mostly quite difficult to ascertain. We will keep our search limited to a dozen thinkers about which we could find information and an interest in communicating the "how" through their writings.

SELECTED THINKERS AND PROBLEM-SOLVERS AND HOW THEY THOUGHT

Da Vinci

A definitive treatise on da Vinci was set forth by Gelb in 1998 [1]. In that work, he documents the "seven da Vinci principles". The "how" of da Vinci is drawn from these principles as below.

71

Curiosity. He maintained a critical curiosity which meant that he pursued his subjects relentlessly.

Refinement of the senses, especially sight. This implies that he was constantly looking and seeing, the latter being from various angles and visualizing

Balanced "whole brain" thinking. He knew how to bring science and art considerations together, possibly in ways never seen before. We might also call that "systems" thinking.

Toleration of ambiguity. He was able to be in that never-never land of uncertainty and lack of clarity. This meant that he was able to keep the targets in sight without giving up. He was able to keep going even though many issues remained unresolved.

Einstein

This author has three biographies of Einstein on his personal bookshelves. Yet it was possible to distill only five ways in which he did the "how" of his thinking. These are [2]:

a. Visualization
b. Combinatory play
c. Intuition
d. Imagination
e. Gedanken experiments

Visualization. This was perhaps the real answer to Einstein's creativity. He worked hard at conjuring up a "picture" of the phenomenon he was investigating. He strove to bring up an image or two, and then stared at it for quite a while. It was as if he was waiting for the image to reveal itself through action and interaction by the movement in the images. It was like watching a movie, if you will? And, in addition we take note of the fact that visualization is one of the new ways of thinking cited in this book.

Combinatory play. This involved bringing disparate pieces together in various unpredictable combinations. What if we put this piece together with this piece? What happens then? What if we keep changing the combinations? What do we see next? We keep "playing" with the combinations. If we had a back pocket computer we would try more combinations even faster. But he did not have such a machine in 1905.

Intuition. Einstein allowed his intuition free-rein. It was all right to have intuitive thinking come upon the scene and appear all of a sudden. He accepted that without having to explain or justify the intrusion. It was like a friend appearing, all of a sudden.

Imagination. Approached more-or-less like intuition. Open the gates to new and even strange thoughts. He was quoted "imagination was even more important than knowledge"

Gedanken Experiments. He often engaged in this practice. He would dream up a situation and explore the ply-through of that situation. Often, they had to do with gravity (so this man is in an elevator going up and he passes a man going down), or they had to do with velocity of travel (so this man is on a train traveling at the speed of light and he sees a man on an adjacent train, also traveling at the speed of light). He then asked, "what does each man see, and experience?" Then he would adjust the parameters. What if...What if...

As a backdrop for the above, to "get him in the right mood" so to speak, Einstein would take a measured break with music and/or art. For the music, at times it was playing the violin. At other times it was listening to someone else on a recording. For art, it often was a trip to the museum to be inspired by his favorite artists who were the likes of Calder, Chagall and the furniture artist Eames.

On the "no" side, Einstein kept away from the purely "rational" approach. He knew that that approach basically did not work for him. One might also call this the "logical" approach. All this appears to be a bit strange, but various books and papers support these positions.

So here is our snapshot of the "how" part of Einstein's approach, boiled down to about a page of text.

Newton

Sir Isaac Newton was one of our great scientists and thinkers. He first set forth the relationships for gravity, which basically held since Einstein [3]. They even hold today in their domain. He was the one who modestly claimed that to the extent his research was important, it was because he was able to stand on the shoulders of geniuses. At the same time, he was combative when he thought such was needed, as in his battles with Liebniz with respect to calculus. He also did significant work in the field of optics. His work on Principia is thought of as one of the greatest ever conceived.

At this point, we seem to have to infer the "how" of what Newton did. In this connection, a best inference here would be to simply say that he was a diligent researcher [3], knew everything that came before him and did use the so-called rational approach. In terms of his method of research, an article points to the intense way he mangled his books, apparently to gain as much as possible from them. Literally, he would squeeze the juice out of them.

Descartes

As we know, it was Descartes [4] that came up with the well-known phrase "I think, therefore I am". This is basically an existential statement, and we accept that Descartes viewed his ability to think as existential. Some think that another way to consider his form of thinking is more like using the word "cognition". Descartes apparently called thinking the mind's principal attribute. His overall ontology consisted of *substances, attributes and modes*. His overall approach is deeply philosophical and leaves little room for a third party to figure out the "how" of his thinking. Do we decline the temptation to do so in this treatise.

Feynman

Richard Feynman was a Nobel Prize winner in the field of physics. He had an expansive personality with many interests in many fields [5]. He was a challenger – challenging himself and his students to take new looks at the world around all of us. He was refreshing, and we would expect that he would reveal some of his "thinking" secrets to all of us.

As it turns out, he seemed not to be withholding when he revealed his major method for his "how" of thinking. It is simply "by analogy". We take note of the fact that this way of thinking is one of the dozen "new" ways of thinking as defined in this book. He points to a few such cases and follows up with another general approach – all theories are under question all the time. That is, one needs to always allow for alternative theories. That is certainly useful for any and all that wonder about the "how" of thinking.

Edison

Of course it was Edison that gave his best advice – he claimed that invention was 1% inspiration, and 99% perspiration [6]. And he knew better than most, having some 1000 patents to his credit as well as several businesses based upon his inventions.

His general way of thinking was looking for quantity of ideas. Literally he gave himself an idea quota and pushed himself to achieve that quota [5]. Numbers cited in this regard are one minor invention every 10 days and a major invention every 6 months. Other citations are that it took over 50,000 experiments to invent the alkaline storage cell battery and 9,000 to perfect the lightbulb. He also systematically challenged assumptions which allowed him to push above and beyond the state-of-the-art. He was meticulous in keeping notebooks so that nothing could be lost – an old idea could always be retrieved.

So the bottom line with Edison is simply that he was a genius to start with and brought intense work habits to the laboratory 7 days a week. He was what today we would call a workaholic. Most would not wish to be working all the time, but for Edison it was not work, it was his way of being. On the other hand, we see that characteristic with many of our "superstars". It was not an imposition – it simply was who they were.

Russell

Sir Bertrand Russell is a renowned philosopher from the UK [7]. He is especially known for his articulation of "critical thinking". His ways of describing this dense subject reveal, in part, how he approaches the matter of how he thinks.

I would cite the following two key suggestions as indicative of how he thinks:

a. Must search for an impartial solution
b. Must identify and question assumptions

He also focuses on establishing useful habits. In particular, here are several of them:

a. The habit of weighing evidence
b. The habit of seeing reality how it is vs. how you imagine it to be
c. The habit of living from one's center
d. The habit of being able to reject hypotheses that prove to be false
e. The habit of maintaining a stance of open-mindedness

Also, we will see others that emphasize "challenging assumptions" as an integral part of how they do their thinking.

Hawking

Steven Hawking is able to give advice to the world, some of which we would also describe the "how" of his approach to his own thinking. Here are four of them [8]:

Never underestimate yourself. Having been stricken with Lou Gehrig's Disease since age 21, it would have been easy for Hawking to give up on being creative. He did not do so, to his immense credit. This must imply that he never underestimated what he could do, despite his disability.

Allow Yourself to Make Mistakes. Only the foolish think that they are infallible. In the scientific world, mistaken hypothesis followed up by a mistaken theory is set forth with great regularity. Forgive yourself if you go off the rails every now and then, so long as you admit where you are.

Be persistent. Stay with the agenda unless there is some fatal flaw in it. Obviously, you would not have achieved what you had if you had abandoned the chase.

Maintain your curiosity. Again, from the position of a terminal illness with great limitations in movement, Hawking could easily conclude that physics research was just too much trouble and required too much effort. It is clear that he did not do that.

Franklin

Benjamin Franklin [9] was the number one self-improvement person. He did it all by himself, and with a fierce determination and discipline. His is a "rags to riches" story whereby he kept at it until he passed at age 84. He diligently read and studied the writings of his favorite authors so this author surmises that such was the method behind his overall "how" of thinking. Keep reading and parsing the works of those he admired, and keep it up for self-improvement.

Franklin is also known for his devotion to four resolutions, and in turn, some thirteen virtues. His disciplined life allowed him to stay committed to these virtues. One can project that they had a lot to do with his overall way of approaching just about everything in his life. And he achieved a lot, from any view or perspective.

Yet one more notion about Franklin is that he was devoted to a holistic existence, and all that such implies. He was truly one of our most well-rounded and productive people of our revolutionary times.

Jefferson

Jefferson was well-known as one of our best thinkers and writers [10]. Some of his quotes reveal how it was that he thought.

The primary was that his thought was through his outstanding skills as a writer. We all know that – how one thinks is demonstrated when we see one's writings. The thoughts are right there, in black and white. In addition, and for general information, he wrote sparsely, and valued the same in others. Here is his quote on that matter:

"The most valuable of all talents is that of never using two words where one will do"

But – he also valued "action", as per another quote:

"Do you want to know who you are? Don't ask. Action will delineate and define you"

Advice to others is very likely to be part of his advice to himself and an important mode of behavior.

"Determine never to be idle...it is wonderful how much may be done, if we are always doing"

Jefferson is clearly a mover and a shaker of his time, especially admired by John F. Kennedy who declared:

"I think this is the most extraordinary collection of talent, of human knowledge, that has been gathered together at the White House, with the possible exception of when Thomas Jefferson dined alone".

Of course, Jefferson was our third President (1801–1807) and had the following short-form accomplishments to his credit:

- Principal author of the Declaration of Independence
- Founded the University of Virginia
- Minister to France and Secretary of State

And it certainly makes sense to pay attention to his quotes, and what they are telling us about him.

Socrates

This takes us back to a key thinker around the 5th century BC. Socrates [11] proposed the Socratic Method which reveals his overall approach to thinking. The Socratic Method involves asking a series of questions which, in the aggregate, defined a way of hypothesis elimination. Assuming that he followed that approach with consistency (he must have) then it means that the "how" for Socrates was to define a large number of hypothesis on a particular matter, follow that up with the questioning dialectic, leading to the elimination of some number of hypotheses. We will not explore the nature of the questions, but we take it as a matter of faith that such a procedure leads to a method of hypotheses elimination. After the elimination, we are left with those hypotheses that have not been rejected.

Saint Augustine

Saint Augustine, otherwise known as Augustine of Hippo, was born in 354 AD in Algeria [12]. His mother was a devout Christian and his father was a pagan who did not come to Christianity until his dying days. His family was Punic, who were Phoenician settlers in northern Africa. Augustine showed signs of a divine and contemplative attitude at an early age.

We have the benefit of many quotes from Saint Augustine, and by their means and inference, we can cite a few of his thinking methods. We start with his saying:

"The punishment of every disordered mind is its own disorder"

From this we are able to conclude that he despised disorder and kept it a long way from his thinking and consciousness. He very likely recognized any tendency in himself toward disorder and rejected it without another thought.

Yet another saying from Augustine is:

"Patience is the companion of wisdom"

Here we have the clue that since he wanted to be (and was, indeed) wise, then he would also be very patient in his approach to thinking. This appears to be simple but is actually an enormous contribution to the engineering and scientific community. How many in these fields have allowed their impatience to stop them in worthy pursuits and possible successes?

Aristotle

Aristotle was born in 384 BC. He is credited with having brought formal logic to his generation and the world at that time [13]. However, in this regard, his work was (and is) so strong that it persisted and is being taught in today's schools.

Here again, we shall benefit from a couple of his quotes to infer what the "how" of his thinking was likely to be. The first quote is:

"Excellence is never an accident. It is always the result of high intention, sincere effort, and intelligent execution. It represents the wise choice of many alternatives – choice, not chance, determines your destiny"

Aristotle is likely suggesting that since a goal of his would be excellence, he would make the definitive choice to go for high intention, sincere effort and intelligent execution. He would, for example, never allow himself to be sloppy in any of these endeavors, at any time or any place. He would be rigorous and careful. He would assure the logic of his arguments. He likely would not allow himself to be playful (as Einstein was likely to do).

Another saying that is worth contemplating is:

"Those who know, do. Those who understand, teach".

I believe that this has been transformed so as to have a different meaning, over the years. My recollection of the transformed version would be "those who know, do, and those who are not able to do, teach". Better in the original, isn't it?

Plato

Plato studied under Socrates, which literally changed his entire life. He was ultimately devoted to philosophy, science and religion [14].

He was born in the year 428/427 BC in and around the city of Athens.

Here again, we are able to infer his areas of thinking from some of his quotes. One of them is simply:

"Don't discourage anyone who continually makes progress, no matter how slowly"

As a teacher, this is a worthy point of view and helpful to his students. But it is also a personal reminder that moving forward is the right answer even if you are stalled for a while.

And here is another quote:

"Human behavior flows from three main sources: desire, emotion and knowledge"

This declaration is somewhat surprising, the more one thinks about it. However, it is his point of view, and he must have thought it through to his satisfaction. Yet another attribution with respect to Plato is that he looked for doing the "right" thing under all circumstance. Doing so would lead to achievement as well as excellence. Not a bad role model for those times, as well as for today.

Darwin

Darwin was born in 1809 in Shrewsbury, Shropshire, United Kingdom, into a well-to-do family [15]. He was well educated and moved on to create the outstanding theory of natural selection and evolution. Here are a few of his quotes from which we will infer some of his thinking.

"a man who dares to waste one hour of time has not discovered the value of life"

This commentary is not unusual and indicates what Darwin most values – his time.

Here are two others:

"my mind seems to have become a kind of machine for grinding general laws out of large collections"

"it is a cursed evil to any man to become as absorbed in any subject as I am in mine"

It appears that Darwin has felt that his devotion to work has been overdone over the years, perhaps leaving the emotional side of his life neglected. Nonetheless, he did not show any tendency toward reform or changing this aspect of his nature. He kept doing what he was doing but these quotes caught him "complaining" about it all. So we are left with this impression – Darwin: a great scientist, complained a lot!

The Bottom Line

So we come to the end of this chapter – one that attempts to elucidate the "how"s of several of our greatest thinkers. This has not been an easy task, trying to figure out how they thought. At times, they made it easy by looking at their writings. At times, we attempted to infer their thinking from their familiar quotations. We urge the reader to go back, page by page and see what the "how" words were. And then we move on.

REFERENCES

1. Gelb, M., *How to Think Like Leonardo da Vinci*, Dell Publishing, New York, 1998.
2. Einstein's Unique Approach to Thinking, see evernote.com/blog.
3. See https://www.inc.com/ilan-mochari/how-isaac-newton-remembered-everything-he-read.html.
4. See plato.stanford.edu/entries/Descartes. Stanford Encyclopedia of Philosophy.
5. Feynman, R., *The Meaning of It All*, Basic Books, New York, 1998.
6. The Edison Method, see link.Springer.com.
7. Russell, S. B., see https://plato.stanford.edu/entries/russell/.
8. Hawking, S., see en.wikipedia.org/wiki/Stephen-Hawking.
9. Franklin, B., see biography.com/scholar/benjamin-franklin.
10. Jefferson; see https://www.monticello.org/thomas-jefferson/brief-biography-of-jefferson/.
11. Socrates; see plato.stanford.edu/entries/socrates.
12. Saint Augustine; see brittanica.com/biography/saint-augustine.
13. Aristotle; see plato.stanford.edu/entries/aristotle.
14. Plato; see plato.stanford.edu/entries/plato.
15. Darwin; see brittanica.com/biography/charles-darwin.

Artificial Intelligence 8

This is a chapter on Artificial Intelligence (AI), which is a field in which machines are able to do many, but not all, of the tasks and functions normally done by the human being. Clearly, computers of various types are a necessary ingredient in this technology. And as computer capabilities have increased, so has the problem-solving strength of AI. The various fields that have been associated with AI have included the following:

a. Machine learning
b. Operating autonomous vehicles
c. Robotics
d. Understanding human speech
e. Natural language processing
f. Massive data processing
g. Artificial neural networks
h. Search methods and algorithms
i. Playing sophisticated games (like Chess and GO)
j. Photo and identity identification
k. Virtual systems

Alan Turing, the mathematician and code-breaker from the UK, set a foundation for AI with his theory of computation, and worked to formulate the Church-Turing thesis which was used as a model for various trains of thought, including artificial neurons and the overall field of neurobiology. Other leading researchers include the familiar names of Marvin Minsky, John McCarthy, Herbert Simon, Allen Newell and Arthur Samuel. Their work and writings have suggested that AI has had its ups and downs over the years, at least as far as government support indicates.

Many of the companies cited in Chapters 2 and 10 have declared their devotion, so to speak, to innovation as a critical capability. Many have linked innovation to AI in the sense that using AI is their approach to achieving innovation. A special report indicates that in 2015, Google was using AI technology for some 2,700 projects within the company.

At least some of the capability of AI was demonstrated by the Watson computer from IBM. In three separate events, Watson played against Boris Spassky (Russian Chess champion), the three Jeopardy champs by the names of Ken Jennings, Brad Rutter and James Holzhauer, and the GO champ by the name of Lee Sedol. In each of those cases, Watson won. So it is clear that a Watson system can be designed and built that will be at least as strong as competing with our very best game players.

THE DoD ARTIFICIAL INTELLIGENCE CENTER

One would expect that the Department of Defense (DoD) would have a keen interest in AI and that indeed is the case. There exists a Joint AI Center (JAIC) with a charter to look across the board within the DoD and try to make sure we are doing all we can in this important arena. In this respect, the JAIC "integrates technology development, with the requisite policies, knowledge, processes and relationships to ensure long-term success and scalability" [1].

The official mission of the JAIC is to "transform the DoD by accelerating the delivery and adoption of AI to achieve mission impact at scale". Note the key words here: accelerate delivery and adoption. In other words, do not develop new AI systems, but where they have already been developed by others, make sure to get them out in the field, solving problems.

The JAIC claims to have a holistic approach to carrying out the following:

- Accelerating the delivery and adoption of AI
- Scaling the impact of AI across the DoD
- Defending the critical infrastructure from a cyberattack
- Enabling decentralized execution and experimentation
- Establishing partnerships in industry and academia
- Assuring a leading AI workforce
- Leading in military ethics and safety

The JAIC carries out its charter and achieves its mission by means of National Mission Initiatives and Component Mission Initiatives. The former are broad and cross-cutting whereas the latter address a very specific issue or problem. The JAIC is considered to be a vital center with a key role in the AI community.

The JAIC and AI Technology

The JAIC has taken the time to explain significant parts of the AI technology in the form of machine learning explanations. There are four families of machine learning algorithms, as cited below:

a. Supervised learning
b. Unsupervised learning
c. Semi-supervised learning
d. Reinforcement learning

Of course it is the human being that does the supervision and deep neural networks can be applied to any and all of the above.

Apart from machine learning is the category of AI called handcrafted knowledge. This category is older. One can see it going back into the 1970s where it was handcrafted knowledge that led to "Deep Blue", an IBM system that played chess against the world's champion and beat him.

The JAIC sees this particular time as a good one for embracing AI technology, leading to improved systems within a company and improved offerings to their various customers. Now is the time to strike, they say, for these reasons:

a. More massive datasets available
b. Greater computing power available
c. New and improved machine learning algorithms available
d. Deeper open source code libraries and frameworks available

Seeing the Forest and the Trees

Our approach here is to examine one application area of AI (from among hundreds) to see what was done and how it was done. This example comes from an article in the *IEEE Spectrum* of recent vintage [2]. It deals with applying machine learning to satellite imagery.

In this forest inventory application, satellite imagery data are being collected at an incredibly high rate. NASA has had an earth resources series of satellites going for some number of years. The current series by the name of Landsat is a three-axis stabilized satellite looking down from a nominal 450 miles. This author, back in the 1960s, worked on an earlier version of this satellite which had the name Nimbus and it was designated as a weather satellite at that time.

The authors of this IEEE paper indicated that they had a "very complicated machine learning" system available for processing remotely sensed forest imagery. For each 1/20 of an acre, the system constructs a list of the trees that are standing there. The list includes features such as species, diameter, tree height, total carbon storage and others. As part of the process, thousands of field measurements were combined with terabytes of satellite imagery. The system provided what was needed to train the machine learning (sub)system to determine the number, size and species of trees represented in the remote-sensed imagery. The team used all kinds of free satellite and aerial imagery, as for example visible light, near infra-red and radar.

The team also claimed to be very fortunate to partner with Microsoft which had a 2017 AI Earth software program. Apparently, this AI software was made available under a partnership arrangement.

This is a worthy example of the use of an AI system to tackle a real-world problem. We presume that this is a representative case and that AI projects are taking place all over this country. For other examples in other domains, the reader is urged to do a google search which remains an excellent source.

Overview of Expert Systems

An important subset of AI systems is the expert system. This type of system emulates the decision-making capability of the human in the system. Typically, it consists of a knowledge base and an inference engine. The former has the subject matter details (about drugs in a pharmacy or Doctor system) and the latter has all the capability needed to access the information in the knowledge base. Some of these expert system applications are described below [3].

Baroid Company's MUDMAN and DEC's XCON. This system analyzes mud during oil drilling.

XCON's eXpert CONfigurer. This system is able to be reconfigured to address a larger sample of issues.

Westinghouse Electric built an expert system to monitor the steam engines that it sells. The system had the name process diagnostic system.

American Express Company used an expert system to allow its credit authorization group to deal with information from thirteen databases. The application in terms of credit is real-time, while the customer is still on the phone.

Coopers and Lybrand built an expert system by the name of ExperTAX to support their accounting staff.

IBM's Consultant Expert System has been effective in helping field service reps price bids for new customers.

Having expert system shells available on the open market is one way to save money and provide an open-source asset. A review of the list of such systems [4] resulted in the following list of expert system shells:

ES/P ADVISOR
Expert-Ease
EXSYS
GURU
INSIGHT 1 AND 2
KDS
KES
M.1
MICRO-PS
PERSONAL COMPUTER PLUS
SeRIES-PC
TIMM-PC

A recommended approach is not to build your own expert system but to review the above list, and other lists, and purchase or lease the one that best suits your needs and gets you into the field with a minimum of expenditure and aggravation.

The State of AI in the Enterprise

A study at Deloitte's [5] addressed the matter of the state of AI in the enterprise. From the executive summary of that study, we find the overall conclusion that:

"Adopters continue to have confidence in the AI technologist's ability to derive value and advantage"

Other conclusions were:

a. The early move advantage may fade soon
b. Essentially all adopters are using AI to improve efficiency; mature adopters are using it to create differentiation
c. Adopters tend to see AI technology as critical to beating the competition
d. Adopters recognize possible risks, but they have mitigated risk by bridging the "preparedness gap"

Another study [6] concluded that the adoption of AI continues to grow in all regions of the world. Over 40% of respondents to our survey said they are currently implementing at least one AI project or plan to do so in the near future. Over 90% plan to implement one or more AI projects in the near future. So here we see a basically positive disposition, but still a reluctance to go all out for AI. This same study suggested a disconnect between the concept of AI and some of the details in such areas as recognition systems, pattern and anomaly detection, predictive analytics and autonomous systems.

The above study revealed some insights into the challenges represented by the adoption of AI. These challenges were perceived to be:

a. Limited AI skills available
b. Data quality or quantity
c. Insufficient ROI justification
d. Confusing AI vendors
e. Non-AI approaches appear to be sufficient
f. Incomplete understanding of AI

Notwithstanding the above, a McKinsey study [7] concludes that the benefits of moving forward with AI outweigh the possible disadvantages. They indicate the following situation:

a. Only a relatively small number of companies have employed AI solutions
b. Early adopters are in a better position to reap the benefits of AI
c. The foundations of AI often lie in "big data", which means little cutting of corners in implementing AI
d. AI brings new challenges, to include new regulatory pressures
   They further suggest that certain factors are necessary for a successful transition to AI, to include:
e. Technology and tools
f. Workflow integration
g. Encouraging wider use of AI
h. Solving the problems of employment and income distribution
i. Considering ethical, legal and regulatory matters

The Salzburg Global Seminar [8] also concludes that AI is changing the world for the better. Areas they cite are:

a. The creation of new jobs
b. The transformation of the government (for the positive)
c. The delivery of better health care
d. Changing and creating art

The latter item is somewhat surprising and certainly raises some questions that need answers.

A Research Agenda

A suggested research agenda from J. P. Morgan [9] has the elements below:

- Data and knowledge
- Learning from experience
- Reasoning and planning
- Human–AI interaction
- Multiagent systems
- Secure and private AI

The stated goal in the finance arena from J. P. Morgan is to explore and advance cutting edge research to develop solutions that will be most impactful with respect to the field in general, and Morgan's clients in particular. We note the first item, a common theme sometimes known as "big data".

The Future of AI

The Office of Science and Technology Policy (OSTP) has decided that what this country does with AI is worth taking a very long and hard look at. They explored "the future of AI" and came to various conclusions, as cited below [10]:

AI and Regulation. Looking to the future of AI, they see new forms of regulation, some of which will be difficult to accept by the populace.

Research and Workforce. They have produced a plan for federally funded R&D containing the elements of (a) the role of Federal R & D, (b) areas of opportunity, (c) ways to coordinate R & D (d) building a highly trained workforce.

Economic Impact. Likely to increase productivity and create new wealth; need to share economic benefits across various sectors.

Fairness, Safety and Governance. Broadening the above item beyond economics.

Global Considerations and Security. One thinks here about international security and the use of AI for advanced weapon systems. New weapons will have astonishing new capabilities.

Preparing for the Future. This report suggests that AI will be a major driver of economic growth and social programs. This will occur only if the three sectors work together, with emphasis on government and industry.

We continue on with a capsule paraphrased version of the twenty-three recommendations from the OSTP regarding the future of AI, as seen in Table 8.1.

TABLE 8.1   Recommendations from the OSTP Regarding the Future of AI [10]

1. Private and public institutions should see how they can responsibly leverage AI and machine learning in ways that will benefit society
2. Federal agencies should prioritize open training data and open data standards in AI
3. The federal government should explore ways to improve the capacity of key agencies to apply AI to their missions
4. The NSTC subcommittee should develop a community of practice for AI practitioners across government
5. Agencies should draw on appropriate technical expertise at the senior level when setting regulatory policy for AI-enabled products
6. Agencies should use the full range of personnel assignment and exchange models to foster a federal workforce with more diverse perspectives on the current state of technology
7. The Department of Transportation should work with industry and researchers on ways to increase sharing of data for safety, research and other purposes
8. The U.S. Government should invest in developing and implementing an advance and automated air traffic management system that is highly scalable, and can fully accommodate autonomous and piloted aircraft alike
9. The Department of Transportation should continue to develop an evolving framework for regulation to enable the safe integration of fully automated vehicles and UAS, including novel vehicle designs, into the transportation system
10. The NSTC subcommittee on machine learning and AI should monitor developments in AI and report regularly to senior Administration leadership about the status of AI, especially with regard to milestones
11. The Government should monitor the status of AI in other countries, especially with respect to milestones
12. Industry should work with Government to keep Government updated on the general progress of AI in industry, including the likelihood of milestones being reached soon
13. The Federal Government should prioritize basic and long-term AI research
14. The NSTC Committee should initiate a study on the AI workforce pipeline in order to develop actions that ensure an appropriate increase in the size, quality and diversity of the workforce, including AI researchers, specialists and users

(Continued)

TABLE 8.1 (Continued)   Recommendations from the OSTP Regarding the Future of AI [10]

15. The Executive Office of the President should publish a follow-on report by the end of this year, to further investigate the effects of AI and automation on the U.S. job market, and outline recommended policy responses
16. Federal agencies that use AI-based systems to make or provide decision support for on cybersecurity, and of cybersecurity on and of these systems, based on evidence-based verification and validation
17. Federal agencies that make grants to state and local governments in support of the use of AI, including foreign government terms of grants, to ensure that AI-based products or services purchased with Federal grant funds produce results in a sufficiently transparent fashion and are supported by evidence of efficacy and fairness
18. Schools and universities should include ethics, and related topics in security, privacy and safety, as an integral part of curricula on AI, machine learning, computer science and data science
19. AI professionals, safety professionals and their professional societies should work together to continue progress toward a mature field of AI safety engineering
20. The U.S. Government should develop a government wide strategy on international engagement related to AI, and develop list of AI topical areas that need international engagement and monitoring
21. The U.S. Government should deepen its engagement with key international stakeholders, including foreign governments, international organizations, industry, academia and others, to exchange information and facilitate collaboration on AI R&D.
22. Agencies' plans and strategies should account for the influence of AI on cybersecurity, and of cybersecurity on AI.
23. The U.S. Government should complete the development of a single, government-wide policy consistent with international humanitarian law, on autonomous and semi-autonomous weapons

Finally, from the OSTP's year one annual report, we provide key elements of the strategic plan for our country, as listed below [11].

1. Invest in AI Research and Development
2. Unleash AI Resources
3. Remove Barriers to AI Innovation
4. Train an All-ready Workforce
5. Promote an International Environment Supportive of American AI Innovation
6. Embrace Trustworthy AI for Government Services and Mission

EXERCISES

1. In three pages, explore the relationship between AI and problem-solving
2. In three pages, how would you use an expert system to (a) enhance your thinking and (b) improve your problem-solving?
3. In three pages, suggest three ways in which the OSTP might improve upon the strategic plan for our country, and not cited in this chapter.

REFERENCES

1. See http://dodcio.defense.gov; joint artificial intelligence center (JAIC).
2. Parisa, J., and M. Nova, "This AI can see the forest and the trees", *IEEE Spectrum*, 57: 32–37, August 2020.
3. Leonard-Barton, D., and J. Sviokia, "Putting expert systems to work", *Harvard Business Review*, 66: 91–98, March 1988.
4. Eisner, H., *Computer-Aided Systems Engineering*, Prentice-Hall, Englewood Cliffs, NJ, 1988.
5. See https://www2.deloitte.com.
6. See https://www.forbes.com.
7. See https://www.alibabacloud.com.
8. See https://www.salzburgglobal.org.
9. See https://www.jpmorgan.com.
10. "The Future of AI", Office of Science and Technology Policy, The White House, Washington, DC.
11. "American AI Initiative, Year One Annual Report", Office of Science and Technology Policy, The White House, Washington, DC.

Group Problem-Solving 9

Groups are often assembled to solve a problem, build a system, share information or reach consensus on a controversial (nontrivial) issue. These groups are often within an enterprise, or from several enterprises. These groups may be within a country, or they may be international.

It takes considerable skill to lead a group, as can be inferred from the above. Yet, many of us are called upon to do so. This short chapter will attempt to help the potential group leader in being effective in this endeavor. We will start out with a quite popular subject, namely, how to do it "wrong". This short section on doing it incorrectly has the subtitle "GroupThink".

GROUPTHINK

This author has explored this topic in a couple of his books [1,2]. GroupThink is a process by which a group tackles a problem and, even after lots of discussion, comes to the wrong conclusion on a particular subject. There are many reasons why this may be the case, but the one most likely is that the group participants do not all participate, or they do so, but they follow the path of another participant, rather than their own path. They may have a view that is contrary to the group conclusion, but they are fearful of expressing it. There are also many reasons for this withholding. But whatever they are, they are not expressed and therefore are not part of the final reckoning. GroupThink is in the category of dysfunctional behavior of a group, and it has been studied and studied, both to understand it better, and also to try to find ways of eliminating it.

An interesting perspective on GroupThink is provided by Carol Dweck in her book on "Mindset" [3]. She points out that GroupThink can occur when the leader is a genius, so that everyone else backs up into the woodwork. Her example is the Bay of Pigs incident. She further goes on to say such was the esteem in which John F. Kennedy was held, and supported by Arthur Schlesinger, the historian. Churchill set up a special group to avoid GroupThink. David Packard gave a special medal to some who, from time to time, disagreed with him. But Lee Iacocca, according to Dweck, encouraged GroupThink by getting rid of people who had dissenting views. So if one is punished for expressing a contrary view, the stage is being set for GroupThink. And from there, we move away from GroupThink and on to what might be considered good or better group behavior.

THE DoD AND GROUP THINKING

One approach is delineated in the Defense A, T and L Journal [4]. This approach recommends starting with a group (team) charter. Such an instrument makes definite the mission, purpose, resources and expected results of the group. The overall processes then suggest a "roles and responsibility" matrix to work from. This is followed by defining processes dealing with communications, decision-making and conflict resolution. Further common ground rules to be followed by all include such items as (a) full participation, (b) do not interrupt, (c) stay on task and (d) no hidden agendas. Examples of ground rules for two acquisition teams are provided in the list below [4]:

GROUP ONE	GROUP TWO
Assume noble intent	Start on time and end on time
Use the power of questions	Do not shoot the messenger
Really listen	One speaker at a time
Stay focused	Everyone is encouraged to participate
Build on each other's question	Come prepared
Respect the perceptions of others	No long-winded speeches
Keep responses short	Collaborate; strive for win-win
Draw others into the discussion	Have fun

The process of converting a group into a team is deliberate and well thought out. The Tuchman model [5] of stages of team development is always on the table. His model consists of the following four elements:

1. Forming
2. Storming

3. Norming
4. Performing

A top performing team can be recognized by its outputs and by its enthusiasm and excitement. Top performing teams are not rigid; they are flexible in terms of process and in terms of accepting new ideas. Top performing teams are built with cooperation and empowerment. Top performing teams reject any element of GroupThink.

Notwithstanding the above, there is such a thing as the military problem-solving process [6]. The well-defined seven steps of this process are provided below:

1. Recognize and define the problem
2. Gather facts and make assumptions
3. Define end states and establish criteria
4. Develop possible solutions
5. Analyze and compare possible solutions
6. Select and implement solution
7. Analyze solution for effectiveness

Although these are generic, it is surprising how often one or the other of these steps is skipped.

We look at the above steps together with the notion that the military supports the notion of "analysis of alternatives" (AoA). This may be considered an elaboration of step four above. Some key aspects of this approach are presented by the Defense Acquisition University. The overall procedure is an evaluation of the performance, operational effectiveness and suitability, and the costs of alternative systems. The recommended study plan for an AoA is provided in the list below [7,8]:

1. Introduction
2. Ground Rules
3. Range of Alternatives
4. Effectiveness Measures
5. Effectiveness Analysis
6. Cost Analysis
7. Cost-Effectiveness Comparisons
8. Organization and Management

The AoA assesses the critical technology elements for each alternative, to include technology maturity, technical risks, technology maturation and demonstration needs. It also might lead to modified set of Key Performance

Parameters of the alternative systems. The overall intention of the AoA has two aspects:

a. To enhance and document decision-making by providing estimates of risk, uncertainty, and relative advantages and disadvantages of the alternatives under consideration
b. To foster joint ownership and demonstrate a better understanding of subsequent decisions through discussions of key assumptions and variables

A more detailed exposition of an AoA, from the example set forth by the DAU, is provided here in the list below [9]:

a. Capability Need, Deficiencies and Opportunities
b. Program Description
c. Threats
d. Operational Environments
e. Operational Concept
f. Operational Requirements
g. Status Quo (Baseline) and Alternatives
h. System Design, Performance and Measures of Effectiveness
i. Life-Cycle Costs of Baseline and Each Alternative
j. Life Cycle Cost per Unit Systems
k. Life Cycle Cost per Specified Quantity of Systems
l. AoA
m. Trade-off Analysis
n. Sensitivity Analysis
o. Conclusions and Recommendations

So, if a group is carrying out an AoA, the above sections provide some guidance as to what such an analysis should contain. Two other excellent sources for those preparing AoA's are listed in Ref. [10].

GROUP DECISION SUPPORT SYSTEMS (SOFTWARE)

There exist many software systems that are available to support group processes. These are called decision support systems. And, by and large, they are advertised as helping a group reach consensus on any given issue. The generic elements of such systems are:

a. A database (knowledge base)
b. A model (decision context)
c. A user interface

Typically, methods that are used in these systems include the following [11]:

Consensus Decision-Making. The group comes to a consensus, and all have agreed to accept that conclusion. This is sometimes called a win-win solution whereby at the end, all participants are basically pleased with the result. You do not expect a minority report from this process and method.

Voting-Based Methods. As implied by the title, actual votes are taken and the answer with the highest average is selected. This produces the lowest Bayesian regret.

Dotmocracy. This provides for large group brainstorming with an unlimited number of trials and ideas.

The Delphi Method. This method, relatively strong in interpersonal communications and listening, has been well researched and well accepted. It involves several rounds of questions, with all participants reviewing the results after each round. They then adjust their answers, if they wish to do so, and the rounds continue until consensus is reached.

Yet another way of looking at Decision Support Systems (DSSs) is by what are called "social decision schemes" which are methods used by a group to combine individual responses to get to a single group decision. Here are five examples of such [11]:

1. Delegation
2. Averaging
3. Plurality
4. Unanimity
5. Random

Commercial Decision Support Systems

There are many commercially available DSSs, with an overview listing below.

- Statgraphics Centurion
- Style Intelligence
- Yonyx

- Ibi
- Qlik Sense
- SAP BusinessObjects
- Wolfram Mathematica
- TIBCO Software
- Salesforce Analytics Cloud
- EIDOS
- Loomio

Each of these, of course, has a different slant on how to support group decisions. Typical slants include business intelligence, risk assessment and management and business performance management. The reader with a need in this area is urged to do some "googling" to explore features and comparisons.

Collaborative Software Systems

We now move into another brand of DSSs, namely, "collaborative" software systems [12]. The difference between DSSs and Collaborative Software Systems is not completely clear, but collaborative systems are worth looking at as a list below of such systems, all of which have been rated, on a top scale of 5.0, at 4.7 or higher.

SYSTEM	RATING (ON A SCALE OF 5.0 MAXIMUM)
Blink	4.8
Airtable	4.7
MPOWR Envision	5.0
Visual Planning	4.8
Bonzai Internet	4.8
Miro	4.8
FileCloud	4.7
Range	4.7
Slab	4.9
Cascade Strategy	4.9
Upwave	4.7
OnBoard	4.7
Project Fusion	4.8
Start Meeting	4.8
Cocoom	5.0

(Continued)

SYSTEM	RATING (ON A SCALE OF 5.0 MAXIMUM)
ReviewStudio	4.8
MeisterTask	4.7
Quire	4.7
Rivers IM	5.0
Process Street	4.7
Wimi	4.7
RevvSales	4.7
Productboard	4.7
Filestage	4.8
ConnectWise Control	4.7

SELF-ORGANIZING AND HIGH-PERFORMANCE GROUPS

Gureckis and Goldstone report on how groups can self-organize into adaptive, problem-solving entities [13]. They claim that they have seen emergent group behavior that "results from the interaction of multiple agents and their environment". They cite some examples where the information processing at the group level is different from that at the individual level. As we move forward with our understanding of group processes, we are likely to see more of this type of research, i.e., changes in group behavior as the groups adapt through their uninhibited interactions.

Another phrase for these types of teams is "self-directed". In these situations, the presumption is that management has sufficient confidence to allow the self-direction. Three examples of self-directed teams are cited below.

The Manhatttan Project. Here we have Robert Oppenheimer leading the charge to build the first atomic bomb. He was "managed" by General Groves, but someone had to be Oppenheimer's boss. But this author believes that Oppenheimer was given lots of room to do what he thought was appropriate, both from a management and a technical point of view.

The Nuclear Navy. In this case, it is a whole lot more than a team. It is the whole portion of the Navy that deals with nuclear reactors and submarines. Admiral Rickover was a legend in this domain, and he reigned supreme. Nominally, he "reported" to the Navy's Bureau of Ships (one dimension of the problem), but otherwise his oversight probably came indirectly from Congress and from the CNO (Chief of Naval Operations) and the Joint Chiefs of Staff.

Kelly Johnson and the Skunk Works. This was sacred ground at Lockheed, where Kelly Johnson had his own skunk works team to build new and advanced aircraft. The rules were established by Kelly and he did what he thought was right within the "confines" of the VP in charge of this line of business.

Other examples of classes of teams in this connection are as follows [2]:

a. Empowered Teams
b. Self-Directed Work Teams
c. High-performance Teams
d. Integrated Product Teams

The latter item goes back to the days of William Perry (1995) as the Secretary of Defense. These teams were the means by which the Department of Defense (DoD) carried out "integrated product and process development". The latter was an important part of reforming the acquisition process in the DoD at that time. However, guidance on how to form IPTs is still relevant at this time. A particularly interesting document in that connection is the "Rules of the Road – A Guide for Leading Successful Integrated Product Teams" from the DoD. It is clear that considerable time and attention have gone into the development of highly functional teams, especially in industry and government.

EXERCISES

1. Write three pages on how you would defeat GroupThink.
2. In three pages, critique the group decision support systems cited in this chapter.
3. In three pages, explain the management principles/ground rules that Oppenheimer used to manage the Manhattan Project.

REFERENCES

1. Eisner, H., *Thinking: A Guide to Systems Engineering Problem Solving*, CRC Press, Boca Raton, FL, 2019.
2. Eisner, H., "Topics in Systems", Mercury Learning and Information, Dulles, VA, 2013.
3. Dweck, C., *Mindset – The New Psychology of Success*, Ballantine Books, New York, 2006.
4. Defense AT & L, (Acquisition, Technology and Logistics), Washington, DC, Jan–Feb 2012.
5. See https://www.armystudyguide.com; see also https://agilescrumguide.com/blog/files/Use-Tuckmans-Model-of-Team-Dynamics.html.
6. See armystudyguide.com/military-problem-solving.
7. Morrow, P., Analysis of Alternatives, Teaching Note, Defense Acquisition University (DAU), February 2011.
8. Analysis of Alternatives (AoA); see directives, energy.gov, 6/6/2018.
9. Analysis of Alternatives (AoA) Handbook, Office of Aerospace Studies, Hdqts AF HAF, Pentagon Room 5C858, Washington, DC 20330-1070.
10. See https://en.wikipedia.org/wiki/Analysis_of_Alternatives.
11. See https://en.wikipedia.org/wiki/Decision_support_system.
12. Collaborative software; see getapp.com/.
13. Gureckis, T., and R. Goldstone, "Thinking in Groups", see philpapers.org/rec/GURTIG, 2006.

Miscellany

10

This chapter contains a short discussion of a selected miscellany of topics that seemed to not fit in other chapters.

INNOVATION

In Chapter 2 we explored a nonrandom sample of annual reports from several companies, to include:

- Xerox
- Amazon
- Microsoft
- SAS Institute
- Northrop Grumman
- Alphabet (Google)
- Intel
- Lockheed Martin
- Apple
- General Dynamics
- Raytheon
- Leidos

We note that this group of companies chose to highlight innovation in their annual plans, projecting that capability very strongly. Are they really innovative? This author believes that they are. Here is a reiteration of their expressions with respect to innovation [1]:

"we provide our clients with *innovative* solutions…"

"the era of the intelligent cloud and intelligent edge is shaping the next phase of *innovation*…"

"defining what is possible is our purpose, and discovery and *innovation* are at the heart of everything we do…"

"machine learning and artificial intelligence are increasingly driving many of our *innovation*s…"

"our unique ability to reimagine the boundaries of *innovation*…"

"we view *innovation* as the lifeblood of our corporation…"

"bringing the best user experience through its *innovative* hardware, software and services…"

"we have a legacy of *innovation*…"

"we deliver *innovative* technology…"

Now that is an interesting preamble to the subject of innovation. And the facts are that successful companies understand the need for innovation, and new technologies. There is no successful corporate strategy that will sit still and milk that cash cow forever. The innovative competitor is just right around the corner.

INNOVATION IN COLLEGES AND COMPANIES

In the author's previous book on "thinking", a listing of our most innovative colleges was provided. This list is reiterated here, along with an update from the next year (Table 10.1 below)

TABLE 10.1 Comparisons of Most Innovative Colleges

PREVIOUS BOOK [2]	UPDATE [3]
1. Arizona State at Tempe	Arizona State at Tempe
2. Stanford	Georgia State
3. MIT	MIT
4. Georgia State	Georgia Institute of Technology
5. Carnegie Mellon	Stanford
6. Northeastern	Purdue
7. University of Maryland – UMBC	Carnegie Mellon
8. University of Michigan	California Institute of Technology
9. Harvard	Northeastern
10. Duke	University of Maryland – UMBC

Observations of interest include (a) dropping off the top ten list on the part of Michigan, Harvard and Duke; (b) persistent standings of Arizona State, Georgia State, Northeastern and the University of Maryland.

To continue on with an interesting comparison, the lists of most innovative companies are shown in Table 10.2.

Here, we wonder how Apple, Google, IBM and Microsoft fell off the list of top ten. Could it have been something important? Is there a "bump" in the activities of these companies that relates directly to innovation?

Fordham Survey

Yet another perspective is provided by Fordham College in their May 2019 survey, using the **American Innovation Index** [2]. The top ten innovators were recorded as:

1. Apple
2. Honda
3. Weber
4. Toyota
5. Amazon
6. Ikea
7. Google
8. Netflix
9. Navy Federal Credit Union
10. Samsung

TABLE 10.2 The Most Innovative Companies

PREVIOUS BOOK [2]	UPDATE [3] (2018)
1. Apple	ServiceNow
2. Google	Workday
3. Microsoft	SalesForce.com
4. Amazon	Tesla
5. Samsung	Amazon.com
6. Tesla	Netflix
7. Facebook	Incyte
8. IBM	Hindustan Unilever
9. Uber	Naver
10. Alibaba	Facebook

Then we discover there is yet another innovation index known as the **Social Innovation Index** for which the top ten is:

1. Honda
2. Toyota
3. John Deere
4. Ford
5. Navy Federal Credit Union
6. General Motors
7. Aflac
8. Chick-fil-a
9. Trader Joe's
10. General Electric

Now that is quite a list, and it looks like Honda, Toyota and the Navy Federal Credit Union take strong innovation positions on both lists. If you have dealings with the Navy Credit Union are you pleased or not?

AMERICA'S INNOVATION STRATEGY

This author, in his previous book on thinking, provided a strategy for American Innovation. It had six top-level elements, which were [4]

1. Investing in the building blocks of innovation
2. Fueling the engine of private-sector innovation
3. Empowering a nation of innovators
4. Creating quality jobs and lasting economic growth
5. Catalyzing breakthroughs for national priorities
6. Delivering innovative government with and for the people

Each of the above six had three to six subelements, providing substance and clarity to the basic idea. In addition, President Obama, at that time, suggested that there were new opportunities beyond the above, to include:

1. Designing smart regulations to support emerging technologies
2. A 21st century call to serve
3. Harnessing financial innovation for national priorities
4. Increasing the role of "demand-pull" in America's Innovation Strategy

This author has no quarrel with these articulations; they are good answers to the question: "what can and should the government do about innovation? But this author believes that it is time to go a step or two further. Here are four examples of an additional set of worthy activities.

1. Create a massive super-funded STEM program
2. Create a Government-based Council on Innovation
3. Sponsor an Industry-based Council on Innovation
4. Sponsor an Academia-based Council on Innovation

Innovation, from all we can see, is the key to this country's future, especially in areas having to do with solving national as well as global problems (e.g., the climate problem).

PROBLEM-SOLVING STRATEGIES

General problem-solving strategies have been well documented. If implemented across the board, they should influence the nuts and bolts of innovation. A list of some of these strategies is provided below in Table 10.3 [5].

This leaves the analyst with some useful information, but digging more deeply into one or more of the above becomes the order of the day. Another approach is to use the list of twelve new ways of thinking (Chapter 4) as a primer in this connection. Yet another approach is to use the information in some of these chapters, as for example how the great thinkers thought

TABLE 10.3 Selected Problem-Solving Strategies

- Insight
- Abstraction
- Brainstorming
- Analogy
- Divide and conquer
- Hypothesis testing
- Lateral thinking
- Means-ends analysis
- Morphological analysis
- Eight disciplines problem-solving
- TRIZ
- System dynamics

(Chapter 7). Any and all of it becomes a means by which one can solve problems. We note that Chapter 6 points out that there is no "universal" method of problem-solving. That bad news has to be accepted by the problem-solver to remove a possible barrier to getting answers.

VISION 2030

The National Science Board (NSB) [6] has documented its vision of the U.S. science and engineering enterprise which builds upon National Science Foundation (NSF)'s science and engineering indicators. In particular, it identifies threats in the science and engineering arena. The NSB urges a commitment to the following types of actions in regard to these threats:

- A review of NSF's positions and recommendations
- Convening university, industry and state partners to identify best practices as well as barriers to using research results
- Engaging with Congress and the Executive regarding new programs and funding, to include STEM-like activities
- Working with the NSF to develop and expand strategies and partnerships, to include international programs

Certainly the NSF is an agency that is in touch with our science and engineering positions, and we need to take them seriously as part of enhancing our status in the next ten years.

DARPA

The Defense Advanced Research Programs Agency (DARPA) was established in 1958 in order to tackle high-risk high-reward technology. The agency itself claims that it is focused on R&D intended to achieve transformative change in distinction to incremental advances. The budget for the Agency in FY2020 was about 3.46 billion dollars. The NSB defines the scope of DARPA [7] as follows:

"research driven by ideas that have the potential to radically change our understanding of an important existing scientific or engineering concept or leading to the creation of a new paradigm or field of science or engineering.

Such research is also characterized by its challenge to current understanding or its pathway to new frontiers"

Among the investments that have been considered successful are:

a. Areas in computer science
b. Areas in telecommunications
c. Areas in military technology
d. Precision-guided munitions
e. Stealth technology
f. Unmanned aerial vehicles
g. Infrared night vision technology
h. The internet
i. The Global Positioning System
j. Automated voice recognition
k. Personal electronics

Due to its basic high-risk high-reward philosophy, there have also been many failures, which are acknowledged by the agency and are expected. As an example, and this might be somewhat controversial, the Agency took on in 2003 the Total Information Awareness program. The objective of this program was to:

"revolutionize the ability of the US to detect, classify and identify foreign terrorists and decipher their plans"

This required the construction of a large database of information that could be mined in order to extract useful and actionable intelligence. This type of charter led to predictable responses from Congress and other groups, and in many ways ended DARPA's activities in the intel community

The success factors identified as critical by DARPA include the following [8]:

a. Limited tenure and the urgency it promotes
b. A sense of mission
c. Trust and autonomy
d. Risk-taking and tolerance of failure

These factors are real in terms of leading to the behavior that DARPA wishes to encourage. The expiration date is printed on everyone's badge so that the sense of urgency is out there for employees to sign up for and for others to see. Another perspective in terms of hiring from the above list is "hire the best people and then trust them". It does not get a whole lot better than that for a government agency. There is much outside interest in DARPA as many oversight players are attracted to the notion that we are able to come up with "transformative" technologies. Engineers and Scientists are especially interested as members of Congress.

Finally, there are groups within DARPA that require special note. These are its Intelligence Advanced Research Project Activity and the Special Operation Command Research, Development and Acquisition Center. These are designed to be highly specialized as well as agile in their operations.

THE EIGHT DISCIPLINES

The Ford motor company led the charge on defining eight disciplines [9] used to approach and solve problems within the company. These were called 8D and later changed to Global 8D. They became a standard in the automotive field, and compared favorably with FMEA, or failure modes and effects analysis. In other words, they began to look like quality assurance methods, not a terrible thing. The specific eight disciplines are:

1. Preparation and emergency response actions
2. Use a team
3. Describe the problem
4. Develop interim containment plan
5. Determine and verify root causes and escape points
6. Verify permanent corrections for problem will resolve for the customer
7. Define and implement corrective actions
8. Congratulate the main contributors to your team

At the Ford Motor Company these eight disciplines were set forth in a manual called "Team Oriented Problem Solving", or TOPS. Various tools of analysis can and should be considered when using TOPS, namely:

a. Ishikawa (Fishbone) diagrams
b. Pareto diagrams
c. Statistical process charts
d. Histograms
e. Flowcharts
f. Process maps
g. Plan-Do-Check-Act sequences

These items seem overly simple compared to the various disciplines of systems engineering, but they proved themselves to be useful, which is the final test of a suggested method.

THE LEMELSON CENTER

This Center is devoted to the study of invention and innovation. Its stated mission is [10]:

"to engage, educate and empower the public to participate in technological, economic and social change. They undertake historical research, develop educational initiatives, create exhibitions, and host public programming that advance new perspectives on invention and innovation and foster interactions between the public and inventors"

They indicate that their vision is to have a world where everyone is inventive and inspired to contribute to innovation. They are located in the Smithsonian National Museum of American history and their strategic plan of 2016–2020 shifts the Center's mission toward engagement, education and empowerment.

MILITARY DESIGN THINKING

In Chapter 4, we concluded that military problem-solving should be included in our basic list of a dozen ways to approach problem-solving. A subset of that item might be called "the DoD copies industry". In particular, we have discovered that at least some part of the military has accepted "design thinking and problem-solving" as an approach that has merit [11,12]. If the military is adopting some practices from industry, we are following a course that has been used before. An example is the blockbuster initiative back in the 1990s known as Peters and Waterman's "In Search of Excellence". This author can recall the head of an important agency that declared that such was the way of the future. And, in several ways, he was right. So it might well be time for everyone to study and master the design by change approach pioneered by Tim Brown and Tom Kelley.

APPRECIATIVE INQUIRY

There is an approach to problem-solving that fits in the category of qualitative (vs. quantitative) analysis. This has been recorded by the name of appreciative inquiry (AI). It has been described as [13]:

"a collective inquiry into the best of what is, in order to imagine what could be, followed by collective design of a desired future state that is compelling, and thus does not require the use of incentives, coercion, or persuasion for planned change to occur".

We note the focus on a future that one wishes to attain, with this approach being the way to get there. Although it has been contrasted with "problem solving", it certainly appears to this author to be a solving a problem. It has five principles associated with it, as below [14]:

1. **Constructionist**. Our thoughts and actions are derived from our relationships
2. **Simultaneity**. Our inquiry into the behavior of humans changes us
3. **Poetic.** Our behavior in our organizations is expressed by the stories that we tell each other
4. **Anticipatory**. What we think and do today is guided by our vision of the future
5. **Positivity**. Sustainable change requires positive relationships between people

Finally, AI has been thought of in terms of the field of Organizational Development and as a change agent that is both a theory and a practice for organizational transformation.

GROUNDED THEORY

This theory is derived from a situation in which you have a large amount of data on some subject (as for example data obtained from hospital records) and you wish to figure out if there is a theory embedded in the data. The data include both numerics as well as activity-oriented, so that grounded theory may be in part and wholly qualitative. The acknowledged steps in coming to a grounded theory are [15]:

- Search out key points from the data
- Derive "codes" from the key points
- Develop "concepts" from the codes
- Formulate categories from the concepts
- Accept a collection of categories as the definition of the grounded theory

Glaser and Strauss, leading researchers in this field [16], worked together on dying hospital patients to write "Awareness of Dying" in 1965, from which they formulated the "constant comparative method" which became the grounded theory approach. They also commented that:

- The theory fills a gap between conventional theory and empirical research
- The theory develops the overall logic of grounded theories
- The theory legitimizes true quantitative research

As a footnote to this entry, this author found that grounded theory became acceptable as a doctoral dissertation approach during his tenure as a professor.

MORPHOLOGICAL ANALYSIS

This is also a problem-solving technique especially suited for qualitative (vs. quantitative) considerations. It was formulated by Franz Zwicky, an astrophysicist working at the California Institute of Technology [17,18]. It was developed to handle multidimensional problems that were not suitable for causal modeling or simulation. It tends to work backward from a desired result, focusing on a "grid box" of the domain's dimensions. So, for example, we are attempting to "invent" a new automobile. We look at the key dimensions of this domain, which might be identified as (a) type of engine and (b) type of fuel. In the former case, we start with straight, in-line, Vee, Flat and Wankel. For fuel, we start with conventional gas, diesel, liquefied petroleum, compressed natural gas, ethanol, biodiesel and electric. We now construct a grid box which is a two-dimensional map of the five engines against the seven types of fuel. Then we look at each cell in the grid box and ask about whether or not these are currently in the fleet mix. By considering all possible answers we are able to find, if it exists, a new approach. One can readily see how this kind of approach is suited for the process of invention.

THE 2×2 MATRIX

A method that looks somewhat like the Zwicky idea is embedded in the 2×2 matrix approach [19] but is actually quite different. The 2×2 matrix is a mapping of two key independent variables against each other. The x-axis variable has the values low and high as does the y-axis variable. So the four cells have the values: high-high, low-low, low-high and high-low. This four-cell approach has enormous appeal in that it is simple, easy to understand and tells a story, all by itself.

An example of this approach can be seen by mapping these two variables against each other, task behavior vs. relationship behavior. Low-low for both has the name "delegating". High-high on both scales has the name "selling". Low-high and high-low are "participating" and "telling". The reference in this area covers the subject very well and provides numerous other examples from many aspects of life, but emphasizing some type of enterprise.

SYNECTICS

Synectics is a creative problem-solving process whereby one makes the strange familiar and makes the familiar strange. The word synectics comes from the Greek which means "joining together of different and apparently irrelevant elements" [20]. Much of the background of this word and its implementation lies with a group that was part of the company Arthur D. Little (ADL) in Cambridge, Massachusetts. This group had the main charter of producing inventions for their clients. They have a proven history of success in this endeavor at ADL for about 10 years and then went off on their own. The group membership had varied backgrounds, by design (physics, mechanics, biology, geology, marketing and chemistry). The basic hypotheses under which the group operated were as follows:

1. People can be more creative if they understand the psychology by which they operate.
2. The creative process can be more effective if they allow the emotional component to dominate the intellectual, the irrational being more important than the rational.
3. It is the emotional and irrational points of view that lead to increasing the probability of success.

Project success ultimately had a lot to do with building some type of model of the end results. All the members of the group understood that and constantly moved in that direction. Accepting the above perspectives, much of the synectics process was built around analogies, such as:

a. Personal analogy
b. Direct analogy
c. Symbolic analogy
d. Fantasy analogy

This connects to another of the methods in this book, namely, problem-solving by analogy.

One can see how important group leadership is in synectics. Here are four essential characteristics of the group leader:

1. **Extreme optimism**. Must believe in the group and their abilities
2. **Total grasp**. Must have broad and successful background
3. **Synectics grasp**. Must deeply understand the notion of synectics
4. **Psychical distance**. Must keep psychic distance from the process even though he or she is leading the process

One might say that the bottom line is the perspective that in order to apply the theory correctly, one must be able to play with apparent irrelevancies which help to supply energy for the sometimes arduous task of problem-solving.

PROBLEM-SOLVING AND RE-ENGINEERING

Back in 1993, Hammer and Champy came out with a blockbuster bestseller [21] that said most businesses are doing it wrong. They are getting work done by functional departments instead of process teams. What is needed is a massive change across the board, using their re-engineering the corporation approach. This was problem-solving on a grand scale, and by and large, industry bought into it. And it called for large-scale redesign, emphasizing the following points:

a. Fundamental
b. Radical
c. Dramatic
d. Process-oriented

The orientation to process was critical, and the authors believed that this could not readily be done in-house. It had to be done by people who were trained in the new point of view. But whomever set forth the new way to doing business, it remained a problem-solving exercise. That is, just about everyone has a problem that needed to be solved.

In effect, this book takes a different approach. It is implicitly saying that if a corporation has one or more problems, there are many approaches to its solution, and the enterprise is fully capable, in general, of solving the problem in-house. A team approach is recommended, and a selection of the best approach, from among many, is the way to go. So – do not go off running and tearing the company apart and hiring a bunch of consultants well-versed in process design; choose a problem-solving approach from those suggested here and let your internal high-performance teams find the right solutions.

COST EFFECTIVENESS

Many solutions to company problems take the form of finding the correct answer by means of a deep cost-effectiveness analysis. That is, the step that calls for comparing alternative solutions should be based upon finding the most cost-effective solution. So we take some time here to explore what that might look like.

The cost side of the picture has two basic approaches:

a. Enumeration of costs in context of life cycle cost model (LCCM) [22]
b. Cost estimating relationships (CERs)

The LCCM has three top-level costs, namely:

a. R, D, T and E (research, development, test and evaluation)
b. Procurement
c. O&M (operations and maintenance)

There are close to fifty elements of cost under the above three categories, and 10 years for the LCCM, so that some 500 elements of cost need to be estimated for this approach.

The formal CER approach can be used to estimate some of the cost elements in the LCCM. There is a Cost Estimating Group that can help in this regard [23]. An example of a CER is the Constructive Cost Model (COCOMO) II [24] model developed by Barry Boehm and his team. This is quite advanced version of a CER leading to estimates of software costs.

So one approach is a mixed one whereby both LCCM and CER are used, i.e., the CER results are fed into the LCCM structure.

COST ASSESSMENT DATA ENTERPRISE

In order to assure accurate cost estimates within the DoD, a CADE (Cost Assessment Data Enterprise) has been established [25]. Their purpose is to collect relevant cost data and try to obtain accuracy in the overall Department's cost estimating. To this date, the CADE has produced annual cost assessments which reveal what has been done, how it has been done and various details of the CADE answers. They also report the existence of a network-based enterprise-level data system for Operations and Support cost information system. The CADE indicates that "the guiding vision for this work is the need for independent, rigorous, and objective cost and schedule estimates paired with thorough assessment of risk, based on solid analytic methods".

MEASURES OF EFFECTIVENESS

On this side of the cost-effectiveness procedure are the specific measures (measures of effectiveness (MOEs)) which are often used as evaluation criteria. Examples of such measures for a Communications System and a Transportation System are provided below [26].

MOEs for a Communications System

- Availability
- Bandwidth
- Capacity
- Grade of Service
- Number of Channels (by type)
- Quality of Service
- Reliability
- Response Time
- Security
- Speed of Service

MOEs for a Transportation System

- Availability
- Capacity
- Comfort and Convenience
- Environmental Effects
- Frequency of Service
- Fuel Consumption
- Maintainability
- Quality of Service
- Reliability
- Safety and Security
- Speed
- Trip Time

LEONTIEF MODEL

This model, formulated by Wassily Leontief, deals with the economics between sectors of the national economy and their interdependencies [27]. For Leontief's contribution, he received the Nobel Prize in economics. The model shows relationships within various sector economies, such that outputs from one sector are inputs to another sector. All sectors are accounted for in the model. Each column of the input-output matrix represents inputs to each sector and the rows represent the values of each sector's outputs.

This large-scale economics modeling was considered a breakthrough in the field and is used today to establish certain aspects of national economics planning. It can also be used for different regional economies.

COMMENTARY ON SYSTEMS OF SYSTEMS

Considerable attention has been paid, in the systems engineering community, to problem-solving with respect to systems of systems. These are usually very large systems in which the interactions between subordinate systems are plentiful. A good example of such a system is the National Aviation System (NAS). This author had a unique experience in terms of the NAS in that he had a contract with the FAA to formulate a model of the NAS back in 1973. Here are some of the "lessons learned" from that activity.

1. Try to think in terms of interconnected, but separate, models
2. Pay special attention to the various stakeholders and how to deal with each of them
3. Do not try to optimize the entire model; deal with each submodel on its own terms
4. In relation to the above, resist the temptation to replace old technology with new technology unless the latter is thoroughly proven

PROBLEM-SOLVING BY REDUCED CLOCK SPEED

Charles Fine, a professor at MIT's Sloan School of management, studied and documented a variety of enterprises and came to some interesting conclusions that are pertinent to problem-solving and new ways of thinking [28]. Here are some of his findings:

a. Clock speed is a significant company parameter that needs constant measurement and working upon.
b. Clock speeds are the rates at which an industry evolves in terms of product, process and/or organizational change.
c. Various industries have different clock speeds which can be measured and documented.
d. Typical clock speeds are rapid ones (in computers and entertainment) and slow clock speeds (aircraft and automobiles).
e. If you are in an industry with a slow clock speed, you may find it advantageous to work at increasing your clock speed so as to pull ahead of your competitors.
f. A first place to look in order to reduce clock speed is supply chain process and management

The reader may wish to look more deeply at Professor Fine's detailed suggestions in this domain.

GEORGE WASHINGTON UNIVERSITY CLASS USING "THINKING" AS TEXTBOOK

The author's book on thinking is an important source for developing problem-solving skills. Its first use as a textbook is at the George Washington University in the School of Engineering and Applied Science. The curriculum from that course is cited below.

Course Title: Systems Analysis and Management 1

Instructor: Dr. Thomas H. Holzer, D.Sc.

Day and Time: Mondays at 6:30 pm–9:50 pm, 3 credit hours

Course Description: The systems or holistic approach as a methodology for making decisions and allocating resources. Analysis by means of objectives, alternatives, models, criteria and feedback

Textbook: Eisner, H., "Thinking – A Guide for systems engineering problem-solving", CRC Press, 2019

Courses Meet via Blackboard Collaborate Ultra

Learning Objectives. At the completion of this course the student will be able to:

1. Identify the major stakeholders of a problem situation or improvement opportunity
2. Develop a conceptual model of problem/opportunity situation
3. Elicit the primary and subordinate objectives a decision maker is trying to achieve
4. Develop an efficient and effective data collection approach
5. Learn the application of Functional Description Diagrams, Causal Relationship Diagrams, Data Flow Diagrams, Behavior Over Time Graphs and the Systems Analyst Worksheet
6. Develop feasible solution alternatives
7. Analyze the solution alternatives to recommend an implementable approach
8. Document and communicate the proposed solution

CLASS	TOPIC/ACTIVITY	ASSIGNMENT DUE
Class 1	Course intro; intro to systems concepts	No assignment
Class 2	Problem definition; problem solving (Part 1)	THINKING, Chapters 1 and 2; another intro to SSM; Homework 1
Class 3	Problem definition; problem solving (Part 2)	THINKING, Chapter 3; Homework 2
Class 4	Data acquisition	THINKING, Chapter 4 Homework 3
Class 5	Analysis approach Mid-term review	Mid-term Homework 4
Class 6	Development of Alternative Solutions (Part 1)	THINKING, Chapter 5 Homework 5
Class 7	Development of Alternative Solutions (Part 2)	THINKING, Chapter 6 Reflective Practitioner 1 Homework 6
Class 8	Implementation Planning	THINKING, Chapters 7–9 Reflective Practitioner 2 and 3 Discussion Board Report
Class 9	Write Final Report; Systems Analysis & Management Review	Final Exam

EXERCISES

1. Write a three-page overview of the activities of DARPA, different from those in the text
2. Write a three-page overview of the design by change approach, different from those in the text
3. Write a four-page approach to (a) the AI approach and (b) the grounded theory approach (2 pages each)

REFERENCES

1. See chapter two
2. See https://news.fordham.edu/for-the-press/pressreleases/consumers-pick-americas-most-innovative-companies/.

3. See Usnews.com/best-colleges.
4. Eisner, H., *Thinking – A Guide to Systems Engineering Problem-Solving*, CRC Press, Boca Raton, FL, 2019.
5. See "Problem Solving"; https://en.wikipedia.org.
6. "Vision 2030", National Science Board; see http://www.nsf.gov.
7. DARPA – Overview and Issues for Congress - Congressional Research Service.
8. "Innovation at DARPA", DARPA, July 2016.
9. "Eight Disciplines Problem Solving", see https://en.wikipedia.org.
10. Lemelson Center for the Study of Invention and Innovation, at the Smithsonian National Museum of American History.
11. Military Design; https://medium.com/@aaronpjackson/a-brief-history-of-military-design-thinking-b27ba9571b89.
12. Brown, T., *Design by Change*, Harper Collins, New York, 2019.
13. See https://en.wikipedia.org; appreciative inquiry
14. Watkins, J., and B. Mohr, *Appreciative Inquiry*, Jossey-Bass, San Francisco, CA, 2001.
15. See https://en.wikipedia.org; Grounded Theory.
16. Glaser, B., and A. Strauss, "The Discovery of Grounded Theory"
17. See https://en.wikipedia.org; Morphological Analysis
18. Zwicky, F., *Discovery, Invention, Research through the Morphological Approach*, McMillan, New York, 1969.
19. Lowy, A., and P. Hood, *"The Power of the 2×2 Matrix"*, Jossey-Bass, San Francisco, CA, 2004.
20. Gordon, W., *Synectics*, Harper & Row, New York, 1961.
21. Hammer, M., and J. Champy, *Reengineering the Corporation*, HarperCollins, New York, 1993.
22. Eisner, H., *Computer-Aided Systems Engineering*, Prentice-Hall, Englewood Cliffs, NJ, 1988.
23. International Cost Estimating and Analysis Association, see website at www.iceaaonline.com.
24. Boehm, B., et al., *Software Cost Estimation with COCOMO II*, Prentice-Hall, Englewood Cliffs, NJ, 2000.
25. See cade.osd.mil/about.
26. Eisner, H., *Systems Architecting – Methods and Examples*, CRC Press, Boca Raton, FL, 2020.
27. See https://en.wikipedia.org/wiki/Wassily_Leontief.
28. Fine, C., *ClockSpeed: Winning Industry Control in the Age of Temporary Advantage*, Basic Books, New York, 1998.

Summary

11

This final chapter summarizes the main points in this treatise. This is done by subject, rather than chapter and deals with:

a. Twelve problem-solving approaches
b. Twelve thinking approaches
c. Another twelve thinking approaches

TWELVE PROBLEM-SOLVING APPROACHES

The main suggested approaches to problem solving are as follows [1]:

1. The "N step" disciplines
2. Reductionist /Technical Decomposition
3. Modeling and Simulation
4. Lateral Thinking (de Bono's Lateral Thinking)
5. Total Systems Intervention
6. Generalized/Systems Approach
7. Design (IDEO) Approach (note Department of Defense (DoD) and others)
8. AI-based
9. Definitive Methods: Mathematics- and Statistics-based
10. DoD-suggested
11. Decision Support Systems (Software)
12. Cost-effectiveness Analysis

TWELVE THINKING APPROACHES

The dozen approaches to thinking, drawn from the author's prior book, are as follows [2]:

1. Inductive thinking
2. Deductive thinking
3. Reductionist thinking
4. Out-of-the-box thinking
5. Systems thinking
6. Design thinking
7. Disruptive thinking
8. Lateral thinking
9. Critical thinking
10. Fast and slow thinking
11. Breakthrough thinking
12. Hybrid thinking

ANOTHER TWELVE THINKING APPROACHES [3]

1. By Visualization
2. By Fable
3. By Algorithm
4. By First Principles
5. Value-focused
6. Focused and Diffuse
7. By Mindset
8. By Debating techniques
9. By Analogy
10. By Expert Knowledge
11. By Osborn-Parnes Problem-Solving Process
12. Use in a Group Setting

EXERCISES

1. In three pages, select the "best" of the twelve approaches to problem-solving. Explain your selection.
2. In three pages, select the "best" of the twelve approaches to thinking, drawn from the author's prior book. Explain your selection.
3. In three pages, select the "best" of the twelve approaches to thinking that are new in this book. Explain your selection.

REFERENCES

1. See Chapter 4.
2. Eisner, H., *Thinking – A Guide to Systems Engineering Problem-Solving*, CRC Press, Boca Raton, FL, 2019.
3. See Chapter 4.

Appendix A
A Dozen Additional Ways of Thinking

In this author's previous book on thinking [1], he presented and discussed some dozen ways of thinking. These are reiterated here in the summary chapter. He also reminded the reader that there are other ways of thinking that were not explored or discussed. Those ways, and others, are briefly examined in the Appendix.

Thinking will be defined here as a particular approach to using your brain. Each approach will be called a "way of thinking". We will explore and discuss another dozen ways of thinking in this Appendix. These dozen will be distinctly different from the dozen that are defined and examined in the author's previous book on the subject of thinking. The new dozen are listed below in Table A.1.

TABLE A.1 A Dozen Additional Ways of Thinking (*)

1. By Visualization
2. By Fable
3. By Algorithm
4. By First Principles
5. Value-Focused
6. Focused and Diffuse
7. By Mindset
8. By Debating (Combative) Techniques
9. By the Creative Problem-Solving Osborn-Parnes Process
10. By Analogy
11. Driven by Acronyms
12. In a Group Setting (e.g., Brainstorming)

(*) above and beyond the author's earlier book on "Thinking" [1].

THINKING BY VISUALIZATION

This item is first on our list since the author recalls how important this mode of thinking was to Einstein. That is an input from a rather serious person. Some of this was passive thinking, and some of it, apparently, was more active in the form of gedanken experiments. Whatever the underlying basis, you can imagine Einstein riding on a train or watching the space-time continuum in a diagram or a picture, or from the deep recesses of his brain.

Scott Thorpe reported on one of Einstein's comments as [2]:

"I rarely think in words at all. A thought comes and I may try to express it in words afterwards"

He (Thorpe) goes on to report that "mental pictures played a vital role in Einstein's thinking and that it was important to Einstein to visualize the problem from its own perspective"

For the rest of us mere mortals, it is conjuring up a "picture" of the problem or a part of the problem. So, as to examples:

a. Visualize the sound waves emanating from a source in a sonar system and what it looks like when these waves bounce off an underwater target.
b. Imagine the impact of a missile with an enemy target, and especially what the lethal flux looked like just before impact.
c. Imagine what a "black hole" looks like just before it swallows an object (like an adventuresome person).
d. Imagine what Maxwell's equations look like traveling through three-dimensional space.

This mode of thinking can be brought into many domains, as for example that of playing music. The mind can visualize the music on a written scale and just about read the music through that visualization. The clef is in one's mind and is literally read from there.

The visualization can be extended to sketches, drawings, key words in context, diagrams and any free-form that comes to mind. Sometimes, mind maps are also part of the visualization process. Also accepted are pretty much whatever shows up on paper when the user has a pen or pencil in his or her hand.

THINKING BY FABLE

This mode of thinking was suggested by Russell Ackoff [3]. His claim was that through this mechanism, coming to a problem's solution is likely to be more immediate. The basic question to be posed is:

- In this situation, which of Aesop's fables might apply?

Here are some examples from Prof. Ackoff's book and other general sources (Table A.2):

For the Tortoise and the Hare, the hare is fleet of foot compared to the tortoise, and so he challenges him to a race. Unexpectedly, the tortoise accepts the challenge, and the race is on. Feeling very confident, the hare naps during the race only to find that it is behind at the finish line. So if you are the tortoise, keep in mind that despite the odds, do not give up.

Do we find this type of competition in the real world? Ackoff claims that we do, and so there are reasons to use this fable to help in deciding what to do. In particular, it may be a good decision to believe in the outcome of the fable.

Moving on to the "Ant and the Grasshopper", we find the ant saving for hard times, and the grasshopper not doing so. Do you see situations where this might apply? Perhaps you see in your family a brother or sister not saving for a rainy day but you are following the strategy of the ant. Do you see this being played out, in one way or another, in your company? Is the company spending too much each month and discounting any possible need in the future? If so, is that behavior likely to affect you?

TABLE A.2 Ten Illustrative Fables for Fabulist Thinking

- The Tortoise and the Hare
- The Lion, the Ass and the Fox
- The North Wind and the Sun
- The Ant and the Grasshopper
- The Crow and the Pitcher
- The Dog and the Shadow
- The Bell and the Cat
- The Wolf and the Crane
- The Lion and the Mouse
- The Boy Who Cried Wolf

For the "Bell and the Cat", a group of mice decide that the invasion of the cat into their territory is to bell the cat so that all know when it is coming. However, there is the delicious question – who will bell the cat? So we have a solution, but it is too difficult to carry out. Here again, do you run into situations that are just too difficult to execute? How about hiring a key competitor? How about pushing beyond the state of the art to gain the advantage that you need (or think you need). In relation to your competitors?

For the "Lion and the Mouse", the lion is about to eat the mouse when it pleads – please do not eat me – perhaps in the future I will be able to help you when you are in need. The lion sees this as a remote possibility and so does not eat the mouse. Lo and behold – the lion gets caught in a net, and the gnawing of the mouse allows the lion to escape. At times, it pays to think about treating others especially well, even when there appears to be no reason to do so. But reasons, however remote, do appear. Can you think of reasons why your company should treat another (smaller) company especially well? Can you help them get new business which they appear to need. Can you help them solve some of their problems? Do you want a special relationship with this company? Do you have a reason for this attitude?

Ackoff believes that there are a sufficient number or situations that would warrant a deeper familiarity with the world of fables. In fact, a list of fables and their meanings could go a long way in terms of helping with problem-solving and decision-making in the marketplace.

THINKING BY ALGORITHM [4]

This method of thinking assumes that one is creating a new algorithm, or one is following an algorithm that already exists. In the former case, new thinking is required. The nature of that thinking depends upon the nature of the problem that is posed. In the latter case, one is simply following a prescribed set of steps and little actual new thinking is part of the process. I had a couple of conversations with my twin grandchildren and they told me they were taking a strange course with the (perhaps) even stranger name of "algorithms". I asked them to explain to me what the course was about. They gave a simple answer – it is about problem-solving and writing the steps down – in sequence. At times, the algorithm can look just like a procedure in a computer program in Basic or Pascal. Certainly, these procedures are part of solving a problem.

Types of Algorithms

From the above text, an algorithm is a procedure, usually a series of steps, that solves a problem. If we look it up from a college course, we find that it appears as a computer science class. And there, a definition is:

- This course presents an introduction to the techniques for designing efficient computer algorithms and analyzing their running times. General topics include asymptotics, solving recurrences and summations, algorithmic design techniques, analysis of data structures, and introduction to NP-completeness.

Various folks have taken the time to categorize types of algorithms [5], with many of these types listed below. We see what a rich topic this is in relation to various forms of thinking.

a. Combinatorial algorithms
b. Optimization algorithms
c. Numerical (mathematical) algorithms
d. Database algorithms
e. Distributed system algorithms
f. Operating systems algorithms
g. Graph algorithms
h. Network flow and theory algorithms
i. Sequence algorithms
j. Search algorithms
k. Kamarkar's algorithm

So we can see that there is no dearth of sources of algorithms that can be accessed to address a variety of problems. Many of these algorithms can be downloaded from convenient websites at no cost or a modest cost. The "thinking" that went into the development of these algorithms may be minimal, or may be extensive. In most cases, no original thinking is required by the user; to use the given algorithm, just follow the instructions. The last cited name on the above list, Karmarkar's algorithm, was in the domain of linear programming.

THINKING BY FIRST PRINCIPLES [6]

This approach seems to have a current champion – Elon Musk, the billionaire and entrepreneur. He has argued that he has used this method and been largely successful at it. It has three basic steps:

1. Look at the problem and question your assumptions about it
2. Break the problem into fundamental principles (as per physics and engineering)
3. Create one or more solutions, from scratch and fundamental thinking

Sounds simple, but of course it is not. For step one, you want to find and reject questionable assumptions – like the cost of power, the cost of solar cells, the cost of manufacturing and the cost of new vs. old processes. Remember – what it costs today using existing methods may likely not be the same as when you install new processes. A piece of data from Elon Musk claims that with new assumptions he was able to reduce the projected overall cost by a factor of ten. That creates a lot of leverage moving from there to a better solution.

Creating new solutions may be considered the most challenging step. Consider the following: first you deconstruct the problem and the current solution, then you are left with several deconstructed "pieces"; you then put these same pieces together in a different way. Thus you have a different solution. This appears to boil down to two steps:

1. Deconstruct
2. Reconstruct

Here again, sounds simple but is not. Remember that the new solutions have to be proven, validated and compared with the original solutions and assumptions.

Musk demonstrates the use of "first principles" with respect to the building of his company SpaceX. It is most interesting that he had the insight needed to prove his point. But real data from real world flights have proven that he was able to do so. Keep your eye on that man. He seems to know what he is doing.

Other examples set forth with respect to this approach to problem-solving have to do with the cases of inventor Johannes Gutenberg, strategist John Boyd, philosopher Aristotle and mathematician Rene Descartes.

VALUE-FOCUSED THINKING

This original approach was formulated by Ralph Keeney and largely set forth in his book on the subject [7]. Another valuable source is his article in the Sloan Management Review [8] which is obviously more compact than the 400 page original sourcebook.

A first key point in value-focused thinking is that alternatives-focused thinking is "putting the cart before the horse". Keeney claims that values are more fundamental for consideration and therefore should be taken first. Further, it is "hard-thinking" and will lead to better decisions. One is led, in value-based thinking, to a variety of factors for quality decision-making, such as those listed below [8]:

    a. Identifying decision opportunities
    b. Guiding strategic thinking
    c. Guiding information collection
    d. Improving communication
    e. Uncovering hidden objectives
    f. Interconnecting decisions
    g. Creating alternatives
    h. Evaluating alternatives

Keeney also tries to clarify by saying that value-focused thinking has two components: first you decide what you want, and then you try to figure out how to get it. With alternative-based thinking, you figure out what alternatives are available and then "choose the best of the lot". He points out that with values-based thinking, you should wind up closer to what you want. That is his overall experience.

Of special importance in Keeney's method are ways of identifying objectives, Keeney gives us some assistance in this matter with a ten-item list, only five of which are cited below:

- Identifying alternatives
- Predicting consequences
- Identifying goals, constraints and guidelines
- Considering problems and shortcomings
- Determining generic objectives

Keeney is clearly devoted to value-focused thinking and is willing to articulate the following top-level benefits:

- Generating better alternatives
- Being able to identify decision situations that are more appealing than the decision "problems" that you are dealing with

Keeney's article [7] also contains a most instructive discussion of his consulting with a firm (Conflict Management Inc.) that led to important strategic objectives in a hierarchy form. This description should be required reading in terms of (a) values-focused thinking and (b) potential effects upon top management. Certainly, Keeney has produced a seminal work by defining and studying values-based thinking.

FOCUSED AND DIFFUSE THINKING

Professor Barbara Oakley was responsible for setting forth and making more well-known the notion of focused and diffuse thinking [8]. These are two modes of thinking and we can switch between the two, even during the course of a day. We use focused thinking when we need to, i.e., when we take on a subject that basically requires more focus. But when this becomes overwhelming, or simply too much, we back away to diffuse thinking. For the latter, we can be more relaxed, but our minds are still working and engaged. However, the level of intensity is much reduced. Focused is like finding yourself in the racing lane of a swimming meet. Diffused is like swimming leisurely to assure your form, but not trying to win a race.

It is claimed that this type of mental oscillation can be tiring, but important. It is further claimed that the following are good examples of this phenomenon:

a. Kerouac's writing of "On the Road"
b. Many of Stephen King's books
c. Edison's and Dali's use of what has been called micro-naps

This author has tried this approach and finds that it is definitely relaxing and grounding. To perhaps oversimplify, it is basically saying – you cannot be "on" all the time, and you do not want to be "off" all the time, or you will not get anything done. So the natural answer is – find the right combination of both.

THINKING BY MINDSET

This method is based upon the book from Carol Dweck [9]. The idea is that people come to various issues and problems in their lives with a mindset – either a fixed mindset or a growth mindset. Further thinking is importantly influenced by which of the two mindsets a person has adopted. For the fixed mindset, one is inclined to feel that he or she needs to prove yourself over and over. Enough is not enough. On the other hand, with a growth mindset, your qualities are evident and can be improved upon year by year. All is well as this growth perspective leads to better mental health and a greater likelihood of success. You are not worried about it – just do the work and the rest will follow naturally. You have got the best overall mindset.

So in this case we do not have a thinking process that is independent of the a priori mindset. In fact, the mindset establishes which way to go – emphasize great difficulty or emphasize getting around and through just about any and all obstacles. Certainly, this is a world that you would like to inhabit. It does not find life so difficult. It does feel like progress can (and will) be made with relative ease just about every day.

Dr. Dweck has a section in her book that is devoted to growth-mindset leaders in business. In particular, she presents very short but penetrating stories about Jack Welch (from GE), Lou Gerstner (from IBM) and Anne Mulcahy (from Xerox).

Jack Welch. The portrait painted of this leader is one of listening, crediting and nurturing. He himself had a growth mindset (obvious in hindsight). From his own approach to life and business he tried to change the culture at GE to one of a growth mindset. By and large, he was able to do this over a period of years. When he became the president of the company, it was valued at $14 billion. Twenty years down the road, the value had increased to $490 billion. And Welch became a hero, among the most admired people in the business world.

Lou Gerstner. The company (IBM) hired Gerstner to literally rescue it from a culture of smugness and elitism, according to Dr. Dweck. The numbers indicated a fall from grace, with lots of trouble ahead unless something drastic was done. It looked drastic to many at that time since Gerstner knew little about the business machine world, as he has come from American Express and RJR. Again, he himself had a growth-oriented mindset and moved forward to establish IBM's new culture in that direction. Some of his actions: opening new channels of communication, creating new highly functional teams, making a host of new deals with just about all of IBM's customers and following through with great precision. By March of 2002, the stock had increased by 800% and IBM had largely restored its leadership in the business machine sector – not an easy place to compete and make progress. As they say – Gerstnew taught the elephant to dance.

Anne Mulcahy. When she arrived upon the scene at Xerox, it was in debt to the tune of 17 billion dollars, with the stock price having dropped from $63.69 a share to $4.43 a share. But by 3 years later, Xerox had four straight profitable quarters and by 2004, Fortune Magazine declared Mulcahy to be "the hottest turnaround act since Lou Gerstner". Here again, supported by a growth mindset, she emphasized learning, toughness and compassion – in hundreds of separate and definitive actions for change.

Dr. Dweck praises the three above leaders, as well as many others that have brought a growth mindset to their companies. Dr. Dweck makes a convincing point – these leaders have leveraged their individual points of view such that their companies had no place to go but to follow suit. Now that is a good model to follow with a company that is basically failing. Somehow, their cultures had gotten de-railed. Back to a growth orientation, and in a well-conceived and rigorous manner, the situation is ready for change. Here are the new rules – follow them or leave, and executed across the board, with compassion.

THINKING BY DEBATING APPROACH

It is no surprise that there are definitive ways to think in order to do debating in an effective manner. This approach is basically combative, and for good reason [10]. Here are some eight steps to consider when using the "debating" approach.

1. First, listen very carefully
2. Restate what you think your opponent's argument is
3. Decide to refute your opponent's argument, or not
4. If willing to refute, construct argument supported by researched data
5. Be able to state with confidence – this is why you (my opponent) are incorrect and why I am correct
6. Expect refutation (once or more)
7. Final summary

If we back up a few steps, a simpler point of view is suggested in the literature [11]. Keep your mind on only three aspects, namely:

- Know your subject, inside and out
- Keep strict control of your actions and demeanor
- Maintain a steady composure

The overall impression as to your disposition if you take the debating approach is to see you and experience your presence is as **contentious**. This is no surprise, considering the basic ground rule of debating. So the reader is cautioned. Tone it down a little so that you come across more reasoned and more reasonable.

THE OSBORN-PARNES PROCESS

This method of thinking is traceable to Alex Osborn in the 1940s and Sid Parnes in the 1950s, respectively. The procedure is part of a Creative Problem Solving process that consists of four stages [12] as shown in the listing below:

BASIC STAGES	STEPS RELATED TO STAGES
Clarify	Explore the Vision
	Gather Data
	Formulate Challenges
Ideate	Explore Ideas
Develop	Formulate Solutions
Implement	Construct a Plan for Implementation

So we assume there is a problem to be solved. To clarify, we articulate all the key aspects of the problem, which can often be placed in a question form. For each such question, we gather basic data from the real world, wherever we can find it. This is followed by a series of challenges. Then ideas are formulated from these challenges. The ideas are "converted" into solutions (on paper). An implementation plan follows. A very brief example of the above is now cited for a problem faced by the Transportation Department some years ago. The problem, as originally stated, was to assess the potential effects of the use of fleets of SuperSonic Transport (SSTs), assuming no improvements in the technology. It was called the Climatic Impact Assessment Program. It took 2 years to carry out the study; looking back on that experience this author finds that the basic method was that of the Osborn-Parnes process along with reductionism ideas [1].

THINKING BY ANALOGY

So it turns out that the Encyclopedia of Reasoning [13], back in 2013, had some strong feelings to express regarding thinking by analogy. To quote their article, revised in 2019:

"analogical reasoning is fundamental to human thought and, arguably, to some non-human animals as well. Historically, analogical reasoning has played an important, but sometimes mysterious, role in a wide range of problem solving contexts…it has been a distinctive feature of scientific, philosophical and legal reasoning"

Here are some of the examples featured in the encyclopedia:

- Joseph Priestly, a strong scientist in both chemistry and electricity, claimed that analogy was his best guide to all philosophical investigations. It had special heuristic value, and for him was particularly useful for research in AI. He drew specific analogies having to do with an inverse square law force.
- Darwin took himself to be using an analogy between artificial and natural selection.
- Physicists, arguing by analogy, looked for spectral line that exhibited frequency patterns characteristic of a harmonic oscillator.
- Descartes dealt with problems correlating geometry and algebra by recognizing their analogous relationship.

So we have strong support for this method of thinking that goes back a long way. More recent expressions include an article by Gavetti and Rivkin in the Harvard Business Review [14]. They point to several strategic decisions in the business world that were based upon analogies. In particular:

- Andy Grove at Intel decided not to cede cheap microprocessors in the marketplace, referring to U.S. Steel and the rebar market (by analogy). He called the issue "digital rebar".
- Thomas Sternberg asked the question "could we be the Toys R Us of office supplies? (thinking about staples)
- Circuit City uses an analogous setting as it opened CarMax (analogy to consumer electronics).

Just how many analogies are there in business and how many discussed in a typical MBA program? It has to be quite a large number. So if the reader is looking for more examples, go back to a well-researched text from a well-conceived MBA program (Harvard will do).

The above arguments make the points that thinking by analogy often avoids detailed and nonproductive analysis – the analogy tells the whole story. But then again, you have to but into the overall logic of the analogy (there is not a "fatal flaw" which makes it unable to be applied).

DRIVEN BY ACRONYMS

This approach is basically using one list (the list of acronyms) to activate a particular chain of thinking. That is a fairly efficient way to go – kind of like a thinking "kickstarter". In this connection, we have included an Appendix (B) [15] of acronyms that have been used by others and are used extensively in the literature.

So – when a particular problem is on the table, you scan the list of acronyms to see which one looks most like it would be useful. You then proceed with this process, moving up other acronyms if you find it useful to do so. It is not a terrible procedure; it is just that the acronym gets "about half of the credit" for the overall thinking process. But then again, who is counting? The objective is to come up with an overall solution, without the presumed necessity of giving credit, one way or another.

THINKING IN GROUPS

The first thing that comes to mind when seeing this topic is likely to have the negative connotation GroupThink. That is not what is addressed here. GroupThink refers to group behavior that is largely dysfunctional and usually leads to poor results [16]. GroupThink is a set of behaviors that was coined by Prof. Irving Janis and studied by many in terms of (a) the behavior patterns and (b) how it can (and should) be avoided.

The point of view is more like: if you are part of a group, and you want the group to be maximally efficient and functional, what is one of the thoughts you should be entertaining? This can be re-phrased in terms of a team: if you are a member (even the leader) of a team, what are some of your major concerns and thoughts?

We continue on here with a citation that is worthy of note. It is the "FORTH" innovation method, where the acronym stands for: Full Steam Ahead; Observe and Learn; Raise Ideas; Test Ideas; and Homecoming [17]. The basic idea is attributed to Gijs van Wulfen, and the method has been proven to be effective in generating ideas in group scenarios. As an example of the degree of success is that FORTH users with 100 ideas brought 78 into development and 51 into some level of market introduction. So the focus here is innovation and as such needs to compete with a variety of other methods where we have groups trying to achieve innovation. We note that in this author's review (Chapter 2) of the annual reports of various companies, we found that a major theme was innovation. Innovation was also a separate chapter in the author's other book on thinking [14]. However, it is not a major theme in this chapter.

Another well-recognized behavior in groups is that of "brainstorming". That is, or course, a generator of ideas usually is rather free form and accepting of inputs without censorship. In this case, you are looking for new ideas – the more the merrier. You do not care if some of the ideas are off the wall; you are able to pare the list down in a later step in the overall process.

A particular form of "brainstorming" is known as the Charette procedure [18]. This is attributed to the French word "charette" which means carts. It apparently was used by architecture students when they passed their work to other students on carts for purposes of constructive criticism. The Charette procedure evolved in that setting, going back to the 19th century. The procedure, typically, will work with a group of some twenty-five to thirty people that are trying to find solutions to problems. The thirty people may be broken down into, for example, six subgroups of five persons each. Then a "recorder" person (same as the Chair) passes out the problems to be addressed, giving them to the subgroups. To further the example, each subgroup may get one problem (so six problems are in play) to work on. They do that work (problem explication and solution) for a period of time specified by the recorder. We will assume that this period is 3 hours, at which time all results are collected and passed (by cart) to all the other groups, as determined by the Chair. This procedure is followed so that there is maximal involvement according to the overall ground rules, which are:

- Several (many) topics will be discussed
- There is limited time to discuss all topics
- It is assured that every person is contributing (must)

All the ideas are graded, as per the equivalent of A, B and below threshold. The Recorder then leads a discussion of what to do next (action plan), typically with the best ideas.

One can imagine that this procedure is well suited to the front end of strategic planning sessions. One can also see it in the context of "reengineering the corporation" if one is bold enough to have thirty or so people tackle a variety of issues in a relatively short period of time and trusts the participants to "do the right thing".

Yet another form of thinking in groups is described by Guereckis and Goldstone [19]. They have reviewed various behavioral studies from their lab that, in varying degrees, looked at groups interacting in real time to self-organize into adaptive problem-solving group structures. Some of the more interesting aspects of this approach examined "distributed" information availability which noted that new cognitive processes may come to the fore in this group setting. This is considered emergent group behavior (self-organized problem-solving) resulting from the interactions of multiple agents in their environments.

We continue on with a brief look at some of the sage advice from Simon Ramo and what one does and does not do in groups [20]. His book had an angle – that of what to do about meetings. Here are some of Dr. Ramo's key points in this domain:

- Be careful about inviting the trouble-maker to your meeting, for obvious reasons (he calls such a person, the MDRSSA [Multidimensional Real Smart Ass])
- Look for the possibility of having people attending such that there will be serious idea augmentation
- Consider the Chair as Leader, but also as chameleon

So we are at this point, asking what type of thinking is needed in groups, not as the group leader but as any member, and all members, of the group. There is an answer that this author wishes to convey, and to do so in one word – integrate. We wish that any and all members of a group think all they are able to think, in the direction of integrating all comments and interactions to improve the overall effectiveness of the group. In this book, the author gives one word to this type of thinking in a group. How this is to be accomplished is left for the consideration in books to come, and by those who are studying group behavior (past, present and future) from a particular point of view.

REFERENCES

1. Eisner, H., *Thinking – A Guide for Systems Engineering Problem Solving*, CRC Press, Boca Raton, FL, 2019.
2. Thorpe, S., *How to Think Like Einstein*, Sourcebooks, Naperville, IL, 2000.
3. Ackoff, R., *The Art of Problem Solving*, John Wiley, New York, 1978.
4. See en.wikipedia.org/wiki/algorithm.
5. See https://www.khanacademy.org/computing/computer-science/algorithms.
6. Musk, E., see https://jamesclear.com/first-principles; text.
7. Keeney, R., *Value-Focused Thinking*, Harvard University Press, Cambridge, MA, 1992.
8. Keeney, R., Creativity in decision making with value-focused thinking, *Sloan Management Review*, 35:33–41, 1994. Summer.

9. Oakley, B., *Focused and Diffuse Thinking*, coursera.org/lecture.
10. See en.wikipedia.org/wiki/debate.
11. Dweck, C., *Mindset*, Ballantine Books, New York, 2009.
12. See projectbliss.net/Osborn-Parnes.
13. See encyclopedia of reasoning.org/metaphor and analogy.
14. Gavetti and Rivkinsee; https://www.researchgate.net/publication/235790263_On_the_origin_of_strategies.
15. See appendix B, this book.
16. GroupThink, see en.wikipedia.org/wiki/Irving_Janis.
17. FORTH; see HOMEFORTHInnovation.
18. Charette; see expert program management.com/2019/charette-procedure.
19. Geureckis and Goldstone; see https://gureckislab.org/papers/dc2006.pdf.
20. Ramo, S., *Meetings, Meetings and More Meetings*, Bonus Books, Chicago, IL, 2005.

Appendix B
Acronyms

CAP: Cover all possibilities
COAL: Curiosity, openness, acceptance, love
CORE: Communication, organizing, relationship, expectation
DO IT: Define the problem; Open mind; Identify best solution, Transform
GASP: General Activity Simulation Program
GERT: Generalized evaluation and review technique
GRASP: Getting results and solving problems
HALT: Hungry, angry, lonely, tired
KISS: Keep it simple, stupid
PDCA: Plan, do, check, act
PEACE: Pause, exhale, acknowledge, choose, engage
PERT: Program evaluation and review technique
OOD: Object oriented design
OODA: Observe, orient, decide, act
RAIN: Recognize, allow, investigate, nonidentification
RPR: Reflexive, Performance, Reset
SIFT: Sensation, images, feelings, thoughts
SLAM: Simulation Language for Alternative Modeling
SNAPP: Stop, notice, allow, penetrate, prompt
SOBER: Stop, observe, breathe, expand, resolve; Short of being entirely ready
STIC: Stop, take a breath, imagine future consequences, choose
STOP: Stop, take a breath, observe, proceed
SWOT: Strengths, weaknesses, opportunities, threats
TAP: Take a breath, acknowledge, proceed
TOTB: Thinking outside the box

Index

AAC (Aviation Advisory Commission)
  Study 7
Acronyms 32–33, 141
Aerospace Corporation 52–53
Allocation, of requirements 68
Alphabet (Google) 19, 28
Amazon 18, 27
Annual Reports 17
ANSER 54–55
Apple 21, 27
Appreciative Inquiry 109–110
  in the enterprise 85–86
Approaches, to problem solving 35
Aristotle 78–79
Artificial Intelligence 81
  in the enterprise 85–86
  the future of 87–89
Availability 46

Better idea stories 26
Blockbuster 29

Capability 47
Climate assessment 3
  Impact assessment (CIAP) 7
CNA 58
Coin weighing problem 60
Coke 32
Collaborative decision support 96–97
Commercial decision support 95–96
Communication 1
Cost assessment 115
Cost effectiveness analysis 46, 114
Crisis, in IT 4
Cybersecurity 4

DARPA 106–108
Darwin, C. 7, 9–80
Da Vinci, L. 71–72
Decision support systems 45–46
Definitive mathematics 42
Dependability 47
Descartes, R. 74

Design thinking 16
  approach 41
Dr. Deming, and Japan and the U. S. 31
Driven by acronyms 137
DoD suggested 43–44
  artificial intelligence center 82
  and group thinking 92–93

Edison, T. 74
Efficient message coding 63
Eight disciplines 74–78
Einstein, A. 72–73
Energy national labs 57
Expert systems 41–42, 84–85

FAA ATC Radars 6
Facebook 21
Federally funded research and
  development 49
FedEx 29
Feynman, R. 74
Five large-scale problems 2
Focused and diffuse 132
Fordham survey 103
Forest, from the trees 83–84
Fourier transforms 67
Franklin, B. 76

Generic step-wise approach 10
  problem and solutions 29
Getting through college 3
General Dynamics 22
Generalized/systems approach 40–41
Good idea 32
Great thinkers 71
Grounded theory 110–111
Group, problem solving 91
  decision support 94–95
  thinking 91–92
GWU Class 118–119

Hawking, S. 75–76
Hewlett-Packard 5

IBM 26–27, 30
IDA 55
Industry, and the DoD 29
Innovation 101–102
    American strategy 104–105
    revisited 25
Intel 20
Intercity transportation 8
IRL (Interrogation, recoding and location)
    System 6

JAIC 83
Jefferson, T. 76–77
Jobs, S. 31

Kalman filter 61

LaGrange multipliers 67
LaPlace transforms 67
Lateral thinking 38
Learning organization 14–15
Leasco 31–32
Leidos 22–23
Lemelson Center 109
Leontief Model 116
Lockheed Martin 21

Mallard battlefield communications 7
Matchstick problem 62–63
Measures of Effectiveness (MoEs)
    115–116
Microsoft 18, 30
Military design approach 109
Minimum computer step problem 62
MITRE Corporation 51–52
Modeling and simulation 37–38
Morphological analysis 111

NAS (National Aviation System)
    Model 6
National Aviation System 64–65
Netflix 29
Nimbus satellite 5
Northrop Grumman 9
N-Step Disciplines 35

Optimal search 66–67
Oracle 28

Osborn-Parnes process 135
Out-of-the-box thinking 45

Parameter dependency diagramming 64
Plato 79
Positive attitude 1
Power, of the idea 25
Predatory pricing 2
Problems, author's experience 5
Problem solving, strategies 105–106
Problem solving, and reengineering
    113–114
Problem solving, reduced clock speed 117

Radar submodel 65–66
RAND Corporation 53–54
Raytheon 22
Research agenda 87
Resilience 1
River-crossing problem 63–64
Russell, B. 75

Saint Augustine 78
SAS (Institute) 19, 28
Self-management 1
Self- organizing 97–98
Social innovation 104
Socrates 27
Specific problems and their solutions 59
Statistics-based 42–43
Steinberg, S. 31
Synectics 112–113
Systems engineering, research center
    (SERC) 50
Systems, of systems, commentary on
    116–117
Surface missile system 8

Teamwork 1
Technical decomposition/reductionism 36
Thinking, as a corporate culture 13
    by algorithm 128–129
    by analogy 135–136
    by debating 134
    by fable 127–128
    by first principles 129–130
    by mindset 132–134
    by visualization 126

Thinking, as necessary in
        problem-solving 13
    in groups 137–138
Think tanks 49
Themes 23
Total systems intervention (TSI) 39
Twelve problem solving approaches 121–122
Twelve thinking approaches 122
Two-by-Two matrix 111–112

Vision 2030 106
Value-focused thinking
        130–131

Walk in the park 68
Willingness to learn 1
What business are we in? 30

Xerox 17, 27, 31

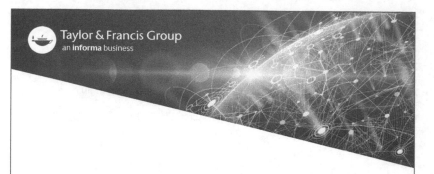